Happiness and Farmers

NIPA® GENX ELECTRONIC RESOURCES & SOLUTIONS P. LTD.
New Delhi-110 034

About the Authors

Prof. (Dr.) Sankar Kr Acharya, former Head, Department of Agriculture Extension and Director, Extension Education, Bidhan Chandra Krishi Viswavidyalaya, Mohanpur, WB, born on 6th October, 1960, started his career as Assistant Professor at BCKV in 1988 and has been in teaching, research and extension over 33 years. An erudite teacher as well as an elegant speaker, Prof. Acharya has also been the topper and awarded the certificate of merit in his M. Sc. (Ag) academic career in Agril. Extension. He is internationally acclaimed for his unique research domain of Social Entropy and Energy Metabolism, Social Ecology and Environmental Sociology, Enterprise Ecology Framework, Farm Stewardships: The Transition from Conservation, Technology Socialization Process, and The Social Ecology of Livelihood, which have got tremendous policy as well as scholastic importance. Research Publication: 235 in National and International Journals. Book Publication: 107 books authored / co-authored, covering areas Participatory Extension Planning and Management; Value, Extension Strategy for Natural Resource Management; Horticulture Management with Quality Issues, Technology Socialization Process; Fish bio-diversity; Income and Livelihood; Enterprise Management and Entrepreneurship Building; Social Metabolism, Farm Energy Metabolism, Community Perception on Climate Change , Social Entropy and Chaos, Decision Ecology Commission; Marketing and Extension Sub Committee, Govt. of West Bengal; Expert member WWF projects in BTR (Buxa Tiger Reserve Project); Expert member; DFID project on Primary Education (DPEP); REIA, Visited Italy, France, Germany, China, Sri Lanka and Bangladesh. Delivered 17 Key note addresses, 13 guest, invited and lead speakers in International Congress, Chaired 19 Sessions in International conferences, Editors of a score of national and international journals. He has so far been awarded with 13 best paper awards in National and International Conferences Organizing Secretary of two International Conferences, held in the years, 2016 & 2018, Jointly Organized by Department of Agricultural Extension, BCKV, WB and Krishisanskriti, New Delhi. His Present RG score: 28.64. He has also been honoured to be selected as the Convener of Panel (PE-32) entitled The hunger , poverty and silence, of IUAES, University of Manchester, UK, 2013 .Co-PI of ICAR, NAHEP and World Bank funded project on Conservation Agriculture. So, far he has successfully guided 19 Ph D scholars and 77 M Sc students from BCKV CU, KU, Vidyasagar University. He is the fellow, ISEE, IARI, New Delhi

Ms Sudipta Chandra, born on 5th November, 1998, completed her B Sc (Hons) Agriculture from Lovely Professional University in 2020 placed in first class. She had her M Sc (Ag) in Agricultural Extension in first class with an OGPA 9.21 which is simply brilliant. She had her M Sc(Ag) dissertation on happiness of farmers under the supervision of Prof S K Acharya, the eminent extension scientist. She has been unique by innovative thinking, mettle and strong academic commitment. She has got a couple of research publications to her credit.

Dr. Amitava Biswas, born in 1964, Professor, Department of Agricultural Extension, BCKV, is an eminent personality in the domain of Extension Science. He has been in teaching, research and extension for the last twenty years in the Department of Agricultural Extension, Bidhan Chandra Krishi Viswavidyalaya (B.C. Agricultural University), Mohanpur, Nadia, West Bengal. He has built up an incredible acumen in Participatory Extension Methods, Indigenous Technical Knowledge, Rural Entrepreneurship and Developmental studies. So far he has authored 45 research papers in national and International Journals, 19 books and good number of popular articles. He is a good trainer in multidiscipline area and regular speaker in TV and Radio. He rendered administrative services as Deputy Director of Extension Education and Editorial Works as Editor of University farm Journal 'Krishi Bidhan'. For the last couple of years he has been with popular Science movement. He has also been involved in some ACRP, NAIP, RKVY and ICARDA Projects.

Mr. Monirul Haque, born on 9th December 1995 in Falakata of Jalpaiguri district presently in Alipurduar district of West Bengal, obtained his B.Sc. (Agriculture) Honours from Palli Siksha Bhavana (Institute of Agriculture), Visva-Bharati, Sriniketan, West Bengal in the year of 2017. He had obtained his M.Sc. (Ag.) in Department of Agricultural Extension from Bidhan Chandra Krishi Viswavidyalaya, Mohanpur, West Bengal in the year of 2019. He has qualified the National Eligibility Test for Assistant Professor conducted by UGC in his very first attempt in the year 2019. Currently, he is pursuing his Ph.D. in the Department of Agricultural Extension of the same University under the supervision of Prof. S K Acharya. He has been awarded Maulana Azad National Fellowship for Minority Students for pursuing his research work.

Happiness and Farmers
The Social Ecology and Strategy

Sankar Kr Acharya
Sudipta Chandra
Amitava Biswas
Monirul Haque

NIPA® GENX ELECTRONIC RESOURCES & SOLUTIONS P. LTD.
New Delhi-110 034

NIPA® GENX ELECTRONIC RESOURCES & SOLUTIONS P. LTD.

101,103, Vikas Surya Plaza, CU Block
L.S.C. Market, Pitam Pura, New Delhi-110 034
Ph : +91 11 27341616, 27341717, 27341718
E-mail: newindiapublishingagency@gmail.com
www: www.nipabooks.com

For customer assistance, please contact
Phone: + 91-11-27 34 17 17
Fax: + 91-11-27 34 16 16
E-Mail: feedbacks@nipabooks.com

ISBN: 978-81-19002-27-6

Composed and Designed by NIPA®.

Prof. J. P. Srivastava
Former Prof. Emeritus
Department of Agricultural Extension
and Communication SHUATS, Naini
Prayagraj, Uttar Pradesh

Foreword

Every man desires to see him happy, and it is impulsive and instinctive as well. Happiness germinates from the seedbed of security, social, economic and ecological securities. Our classical extension research is busy with estimating adoption level since the beginning of green revolution in 70s, and yet it seldom attempts to estimate the level of happiness. The farmers are not the storage house of prescribed technologies by the experts; they are the analyser and inventors as well. Why a farmer should adopts a package of practices recommended if it fails to generate a happy return, the basic minimum to support his family. Indian farmers are possible the only farming Diasporas on this planet who are motivated to go for higher production but no monthly or seasonal income is assured! They are to go for cultivation without market assurance, deterrents to weather aberrations, dedicated monthly income and a minimum market support! For him, before production everything is certain like input, money, technology, advocacy, cost of labour, agro-chemicals, and, after production, almost everything turns oblivious and uncertain like market, income, livelihood and social security. These all system contradictions are certain to go for creating unmageable social entropy. I am really amazed to that the book, Happiness and Farmers: The Social Ecology and Strategy, authored by Prof Sankar Acharya, Sudipta Chandra, A Biswas and Monirul Haque has endeavoured so innovatively and intuitively to elicit the systemic and perceptual factors impacting on farmers' happiness, which has been empirically quantified and analyzed by applying advanced mathematical models and statistical tools. I am really happy to see young scholar like Sudipta Chandra has accepted the challenge to carry out this unique empirical research under the guidance of very eminent Prof Acharya, a pioneering name in the domain of experimental sociology.

Prof. J.P. Srivastava

Preface

The UN declaration on sustainable goals on 2015, seventeen by count, to be achieved by 2030, certainly focuses on ensuring well being of people, especially of rural populace reeling under faster marginalization process. That is why and how the issues of ensuring happiness have earned humongous importance over others. The happiness for deprived class is attainable by ensuring a dedicated income, health and hygiene, minimal level of learning and social security. Happiness is the most adored goal for everyone on this planet and unquestionably every farmer in India. Farmers are wrongly portrayed as a human machine for producing food against some wages or income. Both wages and income are so uncertain for a small and marginal farmer, a happy return remains illusory and a far away dream per se. The main problem of Indian farming is that this is running with increasing stress for every consecutive month or year they have to survive with a market which is uncertain and income which fragile and wage which goes undulating. Globally on happiness-index India's position is just close to the lowest one. In this book, the conceptual heterodoxy of happiness has been well examined. Both the physical and mental well being of the farmers are extremely important for augmenting and sustaining our food security and economic growth. The book creation has been extremely fascinating and voyaging to explore the cause and consequences of unhappiness of farmers, the most neglected humanity, who are supposed to save us from hunger and famine. Huge no of farmers' respondents, perhaps for the first time in agricultural extension in India, to examine how happy they are! The book has got a splendid combination of mind-mathematics-motivation as reflected through a plethora of hard evidences.

Kalyani, West Bengal, India **Authors**

Contents

1

Introduction

Happiness grows at our own firesides, and is not to be picked in strangers' gardens.

– Douglas William Jerrold

Happiness is the barest minimum to any person for survival and moving with a humble vision and mission either. Unfortunately, our farmers are denied to a glow of happiness, and mantled with clouds of agony, depression and confusuion. They are supposed to ensure food security to all, but themselves have to deal with extreme food insecurity, market uncertainty, undulating income, erosion of natural resources and stereotypes of farming occupation. That is how there has been a sharp rise in the migration from farm to off farm occupation.

This is the need of the hour to estimate their happiness. Agriculture is India's backbone, employing half the country's population. However, it has been confronting with several issues, including unfair and non-remunerative market prices, an unstructured supply chain,disaster driven crop failure, and so on. However, an alarming trend in our country shows that farmers are desiring to quit agriculture by large number , 45-60 per cent, and are committing suicides; this is a threat to our long-term agri-development. Do they cultivate because they desire to, or because they have got no other choices either?

What factors do separate farmers who enjoy farming with happiness from those who do not ! Are these factors socio economic or agro ecological by nature only? What about the local environment and the area economy? We can't accept that the population of 125 million will have happiness with no long-term and the sustainable solution. No happiness can be achieved with an ever-increasing social entropy, chaos and bizarre in the market and seamless disruption in climate.

The survey findings are in line with increasing evidence that the farm sector has staged a turnaround of sorts in the last decade. These gains resulting both from higher output growth rates for most crops as well as improved terms of trade for agriculture may have not accrued uniformly. The Centre for Study of Developing Societies (**CSDS**) survey reveals that the proportion of those who like farming is much higher among medium/large farmers (85-86 percent). Also, the satisfaction level with their present condition is 71 percent for farmers in central India, while only 35 percent for their eastern counterparts. Similarly, 78 percent of farmers in West Bengal have an overall negative assessment of their situation, whereas the corresponding ratios are just 16, 22, and 23 percent in Maharashtra, Madhya Pradesh, and Gujarat.

The cost of production in agriculture is increasing incessantly against a stagnation or decline of income from per unit of agricultural investment made by the farmers. How does a farmer go happy amidst hunger poverty and silence! (Chakraborty, Anannya, Acharya, Sankar Kr. in *Hurger, Poverty and Voice,* 2021) And, this is becoming an exponential agony with a constant contribution made by uncertain weather, unpredictable market price, and inaccessible appropriate technology.

Every man thrives on happiness and earns for happiness and this is no exception to a farmer. Happiness is the state of mind and body with which an individual feels good, secure, and agile. It is a state of well-being. Indian farmers are under incessant stress and unhappiness due to the following reasons.

- Return from agriculture is dwindling everyday.
- The farm incomes are seasonal, unpredictable, and cumbersome.
- The general health and nutritional status of farmers, farm women, and farm children are poor, fragile, and un-attended by poor health care support and unaffordable cost of health care.
- The farm production process is mostly dependent on weather and climatic conditions, which offers uncertainty and unpredictability to generate risk and stress.
- Poverty invites conflicts, both at the family and community level, conflicts lead to disintegration and chaos.
- The farm production process itself involves huge drudgery and stress.

The major conclusion one may draw from the survey, along with other ground-level reports, is that the size of landholdings matters a lot today. The reason that small and marginal farmers , who are more preponderant in eastern India,are less happy is that their holdings have become fragmented to the point of becoming unprofitable. These farmers want to exit and they should be encouraged to. The right way to do ıt ıs by creating more non-farm jobs in food processing, construction, manufacturing, and services, while simultaneously promoting an active market in the leasing of land. Many states now do not even permit such leasing, which happens illegally. If farmers with small holdings can legally lease these out with assured resumption of possession at the end of the tenancy contract period, their interests will be fully secured. Basic questions can arises like-Are they really happy? Return per unit area; Are they really happy with the social status they enjoy? Unless these questions are resolved we can't accept that the population of 650 million will have happiness.

Any social science research having heuristics and applications needs a strong axiomatic structure and organized inquiry. The present study looks for generic heard pieces of evidence on the happiness perception of farmers from a kaleidoscopic vision. The crop ecology in which a farmer interacts confronts and concile has got a decisive impact on his/ her profile of happiness. The isolation of dominant factors contributing to state and level of happiness is going to be the expected output of this research. Accordingly, the variables have been carefully selected properly estimated, and methodologically executed for having a comprehensive and dependable milieu of new age research on happiness.

1.1 Rationale of the Study

The happiness of farmers is the crucial area to study. It needs empirical and conceptual innovations. Farmers are under constant stress and depression. When farmers are under stress, productivity suffers. So the psychic ecology of farmers needs to study in a comprehensive manner along with their decision ecology as well. Around three lakhs farmers committed suicide so far. This is the need of the hour to restore their confidence and ensure happiness. That is how a study is required to elicit the means and ways of earning happiness. Occupational migration from agriculture is inviting another threat to food security and to the farmers' well-being as well. So, research on the happiness of farmers can offer an alternative vista to restrain migration.

Happiness is the prime mover for any product performance or any creative work. An unhappy horse can't win the race, an unhappy farmer too can't perform the winner way. The farmers of our country are working day and night for ensuring food security and economic sustainability. They are the providers who offer jobs to 58.8 percent of our population and 17.8 percent of our GDP so their mental health, motivational level and happiness index should run at a higher echelon.

The present research has been placed upon the following rationale:

i. There is a strong need to study the level of happiness of our farmers through a suitable socio-psychological tool and methodology.

ii. There is a need for identifying the marker variables characterizing and contributing to the mental health well being and happiness of the farmers operating in a unique social ecology.

iii. It is also necessary to go for classify farmers based on their happiness level, motivational pursuit, and psychic ecology. The clichés of holding-based classification have got little to do with estimating the mental health or happiness of the farmers.

iv. Whenever the farmers are moving fast and transforming faster, their mental readiness and adaptability merit constant support and elucidation so in-depth research on both process and product indicators appear as an imperative for organizing well-guided research in this domain.

1.2 Research Questions

i. Is it possible to have a benchmark survey on the happiness level of respondents by terms and conditions?

ii. Are there any pool of variables that are functionally impacting the happiness level of farmers?

iii. Does the variable rely while impacting on happiness level, act and interact, and are they really measurable?

iv. Does the empirical study contribute, based on heard evidence, to policy framing as applicable for the selected research local?

1.3 Objectives

i. To conduct a benchmark survey on the overall happiness of respondents' terms and condition assigned in the study.

ii. To isolate and identify the variables impacting on the happiness of farmers, and the predicted character of the study.

iii. To elucidate the level direction and network of interaction amongst and between the set of predictor variables.

iv. Based on the empirical research to extract the policy implications both for micro and macro sociological aspects.

1.4 Limitations

The most important limitations which have been observed in this research are as follows:

i. Benchmarking the status of reigning of 'Farmers Happiness' has really been a difficult task.

ii. There were problems of overlapping of some variables that could have been avoided.

iii. Some new scales could have been better if created.

iv. Since the concept is complex, it was very difficult to get a qualitative response.

v. Nowadays, farmers have become very profession-oriented and hesitate to interact and generate information without expectation. That created barriers sometimes to getting relevant information unleashed on them.

vi. Since most of the questions/stimuli statements concerns the psychic structure and perceptual complexity of the respondents, more time should have been spent on per farmer so that a concrete visualization would have come up.

2

Literature and Citations

This chapter is an extensive survey of all past studies relevant to the field of investigation which gives knowledge of what others have found out in the related field of study. In any scientific investigation or research, a comprehensive presentation of the review of literature is a statutory requirement. Its main functions are to develop a better understanding of the problem to be investigated, delineate a new area of study and avoid unnecessary duplication, decide on the tools and techniques to be adopted including developing some new ones, also to relate the present study with the previous ones by finding out the areas of agreement and disagreement.

The review of literature of the present study is based on the major concepts related to the field of investigation. As direct references on the entire item are not in abundance, certain specific references along with some indirect references have been incorporated in this chapter for the purpose of meaningful use.

2.1 Concept of Farmers' Happiness

Veenhoven (2017) narrated the pursuit of happiness necessitates an understanding of the circumstances for happiness, the issue has gotten a lot of attention throughout the history of Western philosophy. Happiness research has always been a fertile arena for philosophical inquiry. It was unable to test hypotheses on the subject due to a lack of empirical measures of happiness. As a result, the concept of happiness has remained theoretical and unclear.

Ng (2015) stated that life satisfaction is more susceptible to changes in aspiration levels, which reduces the comparability of the indices that arise. Because of a concern for the happiness of others, life satisfaction and preferences can differ from happiness. Happiness is presented as the sole rational final goal in a moral philosophical argument. All proposed caveats to this concept can be explained by their influence on future pleasure or the happiness of others (thus no qualification) or their apparent acceptability can be explained by our flawed rationality. Using happiness instead of life satisfaction (or other concepts), placing down the dividing line of the zero amount of net happiness, using an interpersonally valid unit based on the just perceivable increment of happiness, and using this method in addition to formal methods for large samples are all simple ways to improve the accuracy and interpersonal and inter-temporal comparability of happiness measurement.

Kaczmarek (2017) stated that in psychology, there are two allusions to happiness. To begin with, happiness is a catch-all phrase for a variety of well-being ideas. Secondly, most theorists in the field of emotion psychology regard happiness as a fundamental emotion. Psychological theories of happiness are based on philosophical concepts like hedonic, eudemonia, and meaning in life, as well as utilitarian or moral philosophical traditions.

Senakpon & Dedehouanou (2013) has been evinced that contract farming has got a positive effect on subjective well being.

Tang & Luo(2020) study describes Political participation and local support have a considerable positive impact on farmer contentment. Apart from having a direct impact on happiness, political participation has an indirect impact on happiness through village support, which acts as a buffer between political activity and happiness. Due to farmer heterogeneity, political participation has a greater impact on the happiness of high-income, male, young farmers, whereas village support has a greater impact on the happiness of low-income, male, and older farmers.

Modelmog (1998) narrated the lives of farmers' spouses are more marked by a deliberate disengagement from the consumerist lifestyle of metropolitan culture than by a traditional orientation. Contrary to the homogenization of culture thesis, these farmers' wives want to separate themselves from the urban lifestyle, which is viewed as "loud" and "impersonal," rather than just adhering to it. They even consider themselves fortunate for not having to rely on what metropolitan society has to offer since "thank goodness we have nature." Peace and beauty can be found there. This essay is based on a qualitative investigation of the counterculture and culture of farming wives living in multi-generational households in the Ammerland region of lower Saxony (Modelmog 1994). Over the course of three years, 17 farmwives were questioned.

Nguyen (2007) narrated that the satisfaction that farmers derive from fish farming is not significantly impacted by either revenue per capita or absolute income from fish farming. However, the model results reveal that relative income from fish farming increases the overall likelihood of a farmer being satisfied with their work, proving that relative income—and not absolute money—is related to job happiness. Farmers who participated in AOP's aquaculture extension activities, who had higher income expectations from fish farming, and who have larger pond surfaces compared to total land area all show higher levels of pleasure. On average, older farmers are happier and more satisfied with their fish farming practices.

Liu *et al.* (2020) narrated the factors influencing the happiness of land-lost farmers are evaluated using the rough DEMATEL technique, and fifteen indicators from four dimensions are chosen to develop an evaluation system for their happiness. Government should improve social environment policies, including land acquisition, social security, and employment, to influence other factors and improve the level of happiness of land-losing farmers, according to the results. In China, as more people move into cities, a sizable and unique group of farmers has lost their land. They encounter a variety of demands from work, life, and psychology after leaving the land where they currently reside, and their degree of happiness is typically low. It may result in new social inconsistencies and disputes, which is not good for China's economic development and social stability, and the land-lost farmers risk losing not just their land but also their jobs and inability to assimilate into urban life. Therefore, it has become crucial for the government to investigate and research ways to gauge and enhance the satisfaction of farmers who have lost their land.

Joseph (2004) narrated that the Short Depression-Happiness Scale (SDHS). was composed of six items that consistently scored well across three data sets Three unfavorable and three favorably disposed of items were chosen to preserve the statistical bipolarity of the short scale. The six items can be added up to provide a single score, according to principal components analysis, which supports a single component solution. Over a 2-week period, it was shown that Stephen Joseph et al. had strong test-retest and internal consistency reliability. It was also shown to have strong discriminate validity with existing measures of free-floating anxiety, somatic anxiety, and hysteria, as well as good convergent validity with the full DHS and other recognized measures of sadness and happiness.

Swami (2007) narrated a quick tool for measuring subjective happiness is the Subjective Happiness Scale (Lyubomirsky and Lepper, Social Indicators Research, 46, 137–155, 1999). In a Malaysian community sample of 290 Chinese and 227 Malays, the validity and reliability of the Subjective Happiness Scale's Malay version were examined. The Malay Subjective Happiness Scale had good internal consistency, a unitary structure, and stability over a 30-day period and across ethnic groups, according to the findings. Additionally, the scale's Malay adaptation showed strong convergent validity with single-item happiness scores. The Malay Subjective Happiness Scale's accessibility is anticipated to make it easier to study happiness among Malay-speaking populations.

Cloutier *et al.* (2014) explain the creation of the Sustainable Neighborhoods for Happiness Index (SNHI), a tool to evaluate and compare how successfully various cities, towns, neighborhoods, and communities adopt sustainable practices and how these practices translate to opportunities for locals to pursue happiness. The SNHI is supported by findings from the primary literature and earlier research that point to links between sustainable development elements and greater levels of self-reported happiness. For the generation of the SNHI, nine subsystems of community development were chosen: waste management, energy management, urban planning, food management, business & economic development, buildings & infrastructure, transportation, and community governance. Using information from the SustainLane US City Rankings and the Green City Index (2011), SNHI scores were produced for sixteen US cities (2007). The Sustainable Neighborhoods for Happiness Distribution (SNHD) was created using a method that was later developed in order to plot future SNHI values for comparison. The data for Athens, Georgia, and Ithaca, New York, were then gathered and used to apply the SNHI score technique. According to our findings, Detroit has the lowest SNHI, San Francisco has the highest, Athens is just below, and Ithaca is just above the mean SNHI on the SNHD. Decision-makers, community stakeholders, engineers, developers, architects, planners, and academics can use the SNHI as a special instrument to evaluate the relative progress and happiness of any neighborhood or community.

Zhu *et al.* (2021) narrated utilizing information from the 2016 China Labor-force Dynamics Survey (CLDS), we look at the connection between rural Chinese farmers' levels of satisfaction and their consumer spending. The potential endogenous problem of happiness is dealt with using a two-stage residual inclusion strategy. The empirical findings indicate that a higher degree of happiness is linked to an increase in overall spending. More research demonstrates that spending more on necessities of life, education, and presence is positively and strongly correlated with higher levels of happiness. We also discover that access to the Internet and household income both enhance pleasure and raise consumption. Compared to their urban counterparts, rural households' consumer expenditure is improved more by happiness.

2.2 Farmer's Suicide and the Mental Health

Behere & Bhise (2009) According to their study suicide among farmers is now a universal phenomenon. Studies have identified that farming as one of the most dangerous industries. According to studies conducted in India, the United States, Sri Lanka, Canada, England, and Australia, farming is one of the most hazardous industries, having a higher suicide rate than the overall population. Farmers' suicides have been recorded in India from a variety of sources. Punjab, Maharashtra, Andhra Pradesh, and Kerala are among the states. Research in Maharashtra's Vidarbha region Indebtedness (87 percent) and decline in the economy was linked. Economic position (74 percent) was cited as one of the most important risk factors for suicide. According to a study, the age-adjusted suicide fatality rate is from 17 in 1995 to 53 in 2004

Sengupta (2020) narrated that Farmers/cultivators are defined in the report as those who farm on their property or leased land, with or without the help of agricultural laborers. Surprisingly, landed farmers accounted for 86% of suicides, while landless farmers accounted for the remaining 14%. In India, farmer suicides hold 11.2% of all suicides. In India, farmer suicide is defined as the deliberate taking of one's own life by someone who relies on farming for a living. Farmers commit suicide for a variety of reasons, including monsoon failure, and huge debt burdens.

Sadashiv (2015) stated India has an agrarian economy, for the vast majority of the population; agriculture is their principal source of income. Politicians, industrialists, and manufacturers/producers, on the other hand, have purposefully ignored the qualitative and quantitative development of the same. In all stages of growth, the Indian government has always given priority to the industrial and service sectors, as well as the collateralized agricultural sector. Agriculture has faced numerous issues in the previous two decades as a result of the Indian government's approach. This problem gets more important after 1991, in the era of globalization. After 1991, India's government accepted WTO principles and signed GATTS, which favored wealthy countries such as the United States, the United Kingdom, Japan, and other European countries while harming developing and impoverished countries. These new adjustments had a negative impact.

Parvathamma (2016) narrated several states are falsifying data on farmer suicides The National Crime Records Bureau of India revealed in 2012 that A total of 13,754 farmers have committed suicide. In 2004, the biggest number of farmer suicides was reported, with 18,241 farmers committing suicide. Crops, government policies, public mental health, personal concerns, and family troubles are all challenges that need to be addressed.

2.3 Hunger Poverty and Happiness

Chakraborty & Acharya (2020) narrated in this modern and dynamic world, where man's emperorship is inexorably being established, whether through the creation of Artificial Intelligence or the understanding of the mysteries of the universe, it is humanity's agony that nearly 750 million people do not have enough food, and nearly half of the world's population, 1.3 billion people, live in extreme poverty. More than one billion people worldwide are suffering from the social venom of hunger, and every 3.4 seconds, one hungry child is lost for all time. Food security is defined as the contribution of various aspects such as food availability, food quality, food affordability, and food that is free of social and gender discrimination. The neo-information divide that has emerged as a result of explicit globalization has exacerbated the problem. This segment of the population is still misinformed about information sources and the consequences of being selectively uninformed.

Kumar *et al.* (2020) stated the establishment of targeted social security and public distribution system (PDS) program may have aided in the rapid reduction of poverty among the rural poor.

2.4 Migration and Farmers' Happiness

Mohan *et al.* (2021) narrated during the nationwide lockdown imposed as part of the government's response to the COVID-19 pandemic, the plight of interstate migrant workers in India, who were trapped in overcrowded cities with no means of subsistence and few basic resources, attracted national and worldwide attention. The historical trajectories and political roots of internal migration in India, as well as its interaction with the urban informal labor market and structural factors of precarious work, are explored. The government's response to the pandemic, as well as the disproportionate impact on migrant workers, are both reflections and consequences of a long-standing pattern of neglect and lack of accountability on the part of the federal government toward the employment and living conditions of migrant workers working in precarious conditions.

Srivastava (2021) narrated migrants were heavily affected by India's and other countries' strict lockdown. While migration occurs for a variety of causes and in a variety of forms, it primarily tries to improve the movers' living and employment prospects while also supporting the growth and development of the locations to which they migrate. However, this does not occur without a great deal of effort and expense. Immigration is influenced by patterns of unequal development, demographic changes, wars, and conflicts. Overall, there has been a global trend toward greater mobility, both across countries and within countries, albeit data on many levels is shaky.

2.5 Weather Uncertainty and Happiness

Berry (2011) stated climate change is causing more frequent and severe weather-related disasters, such as droughts, fires, and floods. The majority of what is known about climate change's potential implications on rural mental health is based on extended drought. But, even though drought is widely recognized as a large and widespread stressor, data is equivocal. Drought and other weather-related disasters may damage the social and economic fabric over time, similar to drought. The economic foundations, on which farming communities rely, and mental health vulnerability in rural areas are extreme. The severity of difficulties is exacerbated by socio-economic hardship. Related circumstances may exacerbate the situation.

Kurukulasuriya & Rosenthal (2003) study the agriculture sector is shaped and driven by a variety of variables. Changes in the domestic and international agricultural markets, as well as market fluctuations policies (such as the type and amount of subsidies, for example (monetary incentives, tariffs, credit lines, and insurance,) management techniques, and contract terms trade, technology kind, and availability, and extension, land-use rules, and biophysical considerations attributes (water resources availability, etc.) pests and diseases, as well as soil quality and carrying capacity illnesses) are among the leading causes of death. Given its inherent connection to the natural world, agricultural production is likewise at a premium. Climate change has put us at the mercy of uncertainties. Extreme events, such as earthquakes, drought, and flooding are examples of the variety. While recent policy debates have centered on reducing emissions that cause climate change, there has been relatively little discussion of other issues. First, both developed and poor countries are already dealing with climate unpredictability. Second, even moderate climate change adds to the urgency of promoting local adaptation choices at the same time as global mitigation initiatives are pursued.

IPCC (2021) Working Group III of the IPCC is in charge of evaluating climate change mitigation - responses and solutions to the threat of severe climate change by reducing emissions and improving sinks of greenhouse gases that cause global warming. Topics covered in the Sixth Assessment Report (AR6) include the relationship between consumption and greenhouse gas emissions, as well as the influence of innovation and technology. The report will examine the relationship between short- to medium-term initiatives and their consistency with the Paris Agreement's long-term temperature goal. It will look at alternatives for mitigation in areas like energy, agriculture, forestry, and land use, as well as buildings, transportation, and industry

Hansra *et al.* (2018) narrated to keep up with climate change in the next years, innovations will be required to increase the rate of adaptation of agriculture and the accompanying socioeconomic system.

2.6 Social Entropy and Relation to Happiness

Acharya *et al.* (2018) narrated the entire Indian farming system is experiencing a complex, polymorphic, and dynamic transition process. Regardless of physical, biological, or social ecology, continuous energy equivalency is required for ecological transition. Matter and energy are the most important movers in physical ecology. Genetics and metabolism are the respective factors in biological ecology; motivation and intelligence are, of course, the extreme determinants in social ecology. Motivations are released in one social system through education, training, simulation, and simulation. These incentives include both genetic and psychological components.

2.7 Psychological Wellbeing

Cox *et al* (2016) narrated a growing understanding of the significance of local community wellbeing indicators as a platform for citizen involvement, community planning, and evidence-based policymaking is reflected in the recent increase in interest in Australia and elsewhere. Indicators of community wellbeing, more generally, are a part of a global movement to reconsider how political goals are discussed and how progress and wellbeing are defined and assessed. This article describes the creation of Community Indicators Victoria (CIV), a regional project for measuring community wellbeing in Victoria, Australia. The CIV program is first described in the article's broader historical and global framework of work on wellbeing, progress, and sustainability indicators, as well as in the context of Australian and Victorian policy.

According to **Sen (2006)** death due to famine is not because of crop failure. At that time soil scientists claim the NPK was fair enough but famine happened because of the 2nd world war. Foods were snatched by the British emperor. 30 lakhs people have died because they didn't have access to food.

Kumar *et al.* (2021) stated the study looks at the farmers' contentment and satisfaction following the land sale for Gurugram, Haryana, India's urban expansion, as part of the "Development Plan 2031AD for Gurgaon-Manesar Urban Complex." Using a convenience-cum-judgmental sample technique, information was gathered from 177 male farmers (168 valid) in sixteen communities whose land had been purchased or sold for urban growth. Factor analysis and PLS-SEM (Partial Least Squares - Structural Equation Modeling) were used to analyze the data. According to the study, income from employment and rental income, as well as increases in spending on children's education, household goods, technology, cars and sport utility vehicles, shopping, leisure, and social events, all have a positive and significant impact on farmers' satisfaction after the land is sold. The aspects of urbanization—urbanized living, nuclear families, and access to essential services like water, power, roads, and transportation—have a big impact on how happy farmers are. The findings also demonstrate that increased living standards and remuneration, as well as their application for long-term financial growth following the land sale, have a significant positive impact on farmers' contentment. Contrarily, societal amenities like banks, hospitals, entertainment centers, educational institutions, and social costs like pollution and health issues have little to no effect on farmers' happiness.

Bhattacharyya *et al.* (2019) narrated in their study that one in five Indians who require mental health counseling have a mental condition that affects less than 5% of them. In India, mental illness will cost the country 38.1 million years of healthy life by 2025. Due to the rising stress of modern life, mental serenity and happiness are now necessities. If we want to stop the ticking time bomb of mental illness from erupting, we need to understand happiness, measure it, and find ways to be happy in our lives. Happiness is characterized by positive or pleasurable emotions, ranging from contentment to great delight, according to psychologists. Humans that are happier have better physiological functioning, are better able to handle stress and discomfort, and have longer life spans.

3

Theory and Concept

This chapter deals with some important terms and their concepts related to the study which are essential to convey the ideas of the study effectively and also to provide a theoretical base for the empirical investigation and guidance for the selection of relevant predictor variables.

3.1 Human Happiness and Estimation

A state of mental or emotional wellbeing known as happiness is characterized by pleasant or good emotions, which can range from contentment to ecstatic delight. Happy emotions can also be a reflection of how someone feels about their general well-being.

The United Nations Sustainable Development Solutions Network publishes the World Happiness Report as a measure of happiness. The UN General Assembly issued a resolution in July 2011 requesting that member nations assess the level of happiness among their citizens and utilize that data to inform public policy. The International Day of Happiness was proclaimed by the UN on March 20 in order to recognize the importance of happiness and wellbeing as universal aspirations. The topic of "happy" has been in the news recently, but India didn't have much to smile about because it was one of the 10 nations that had the biggest drops in happiness between 2005-2007 and 2013-2015. India rated 118th out of 156 countries in the World Happiness Report, down one spot from the previous year. But the fact that Pakistan, Somalia, and Saudi Arabia were ranked as being happier than India must have upset the powers. The World Happiness Index polls a large number of people from a variety of nations each year to determine, as the name suggests, which nation has the happiest citizens. Denmark took first place this year, closely followed by Switzerland, Iceland, and Norway. US performance was 13th.

According to philosophers like Aristotle, the main aim of life for the majority of people is happiness (Frey and Stutzer 2008). According to Bubis and Erceg (2018) and Walsh et al. (2018), happiness is crucial for increasing social stability, job success, and individual decision-making and problem-solving. The nature of happiness and its connection to other socioeconomic factors deserve attention given the significance of happiness in daily living and at work.

For a very long time, the psychology field has focused a lot of its research on the idea of happiness. However, economists did not become aware of it until 1974 (Easterlin 1974). Happiness is a more accurate indicator of one's level of personal well-being than income (Frey and Stutzer 2002). According to a psychological study, affect, especially positive affect, has a significant impact on how people behave. Positive emotions, for instance, can improve people's capacity for creativity, cognition, selection, and altruistic behavior (Oswald et al. 2015). Positive emotions, such as optimism, enthusiasm, helpfulness, and satisfaction, are more prevalent and less abundant in happier people (Isen, 2000). An economic study has recently focused heavily on the impact of happiness on individual behavior, in part as a result of psychological research.

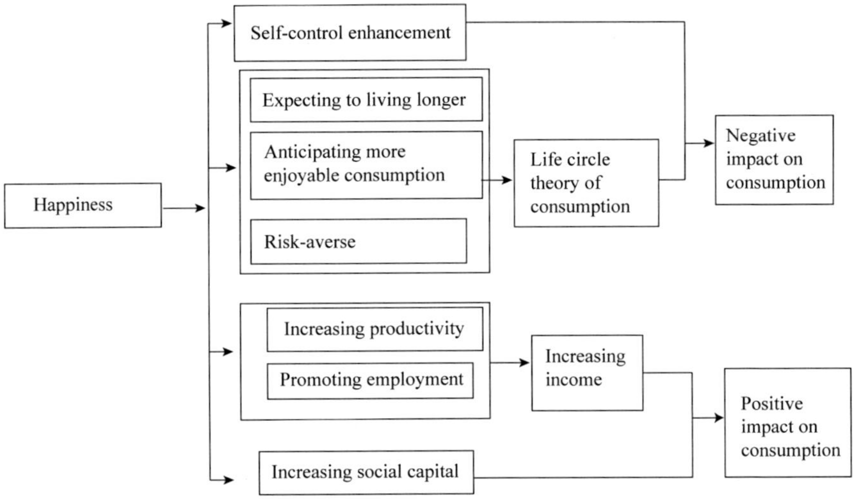

Fig. 3.1: Conceptual framework of happiness-consumption nexus (Zhu et al., 2020)

3.1.1 Intended Uses of the Happiness Index

Researchers, community leaders, and decision-makers can utilize the Happiness Index to better understand and promote community well-being, social justice, economic equality, and environmental sustainability. By making the survey instrument and data freely available to community organizers, educators, researchers, students, organizations, government, and more, the index was created with the intention of fostering social change. The index is distinctive in that it is the only thoroughly thorough index readily accessible for free online that allows survey participants to access their own scores in comparison to the entire data set, while also allowing users to customize the tool for a target population, add their own questions to the survey instrument, and easily access data for their own sample as well as comparison data to the entire data set.

Specific faults accompany these conceptual simplifications. For example, the "availability" heuristic entails focusing on pieces of information that are already in the public domain. The advantage of being in good health is obvious if the interviewer is in a wheelchair. The association between happiness ratings and health indicators will be stronger among respondents who report being in good health. Schwarz and Strack have demonstrated a few of these heuristic effects (1991).

3.1.2 Domains of Happiness

- Psychological Well-Being: optimism, senses of purpose, and accomplishment.
- Health: energy level and ability to perform everyday activities.
- Time Balance: enjoyment, feeling rushed, and sense of leisure.
- Community: the sense of belonging, volunteerism, and sense of safety.
- Social Support: satisfaction with friends and family, feeling loved, and feeling lonely.
- Education, Arts, and Culture: access to cultural and educational events and diversity.
- Environment: access to nature, pollution, and conservation.
- Governance: trust in government, sense of corruption, and competency.
- Material Well-Being: financial security and meeting basic needs.
- Work: compensation, autonomy, and productivity. (Happiness Alliance, 2014c)

Explanations of what it means to be "happy" have been the subject of extensive empirical and theoretical study (see Diener 1984; Eysenck 1990; Myers and Diener 1995; Nettle2005). This body of research has historically concentrated on factors that can be measured and can have an impact on happiness or well-being, such as economic conditions (Juster and Stafford 1985), life events (Headey and Wearing 1989), personality (Costa and McCrae 1984), and level of activity (Csikszentmihalyi 1990). Recent research, however, has revealed that subjective processes play a significant role in happiness and that objective conditions, demographic factors, and life events do not fully explain happiness (Diener 1998).

This subjective method takes into account respondents' own perspectives on happiness. However, due to the difficulty in assessing subjective happiness, researchers have mostly relied on informant data, non-verbal observations, physiological evaluations, and self-reports (see Diener 1994). There are several validated measures for self-reports, including multi-item scales like the Affect Balance Scale (Bradburn 1969) and the Satisfaction With Life Scale (Diener et al. 1985). There are various single-item scales including the Gurin Scale (Gurin et al. 1960), the Delighted-Terrible Scale (Andrews and Withey 1976), the Self-Anchoring Scale (Cantrill 1965), and the Global Happiness Scale (Bradburn 1969).

However, single-item scales measure global judgments that are not suitable for examining psychometric features, whereas multi-item scales analyze diverse aspects of pleasure (such as affective and cognitive states). According to Lyubomirsky and Lepper (1999: 139), a scale that assessed one's total subjective happiness—i.e., whether they are pleased or unhappy—was missing from the literature. They created the Subjective Happiness Scale to fill this gap in the literature, which measures overall well-being through global self-evaluations and reflects a broader and more "molar" category of well-being. Based on data from fourteen distinct samples in the United States and Russia, Lyubomirsky and Lepper (1999) reported that the scale, although very brief, passes basic psychometric standards for measurement accuracy.

The Subjective Happiness Scale was translated into Bahasa Malaysia (Malay), and its validity in a Malay-speaking population is described in the current study. This is significant for a number of reasons. First, the lack of native and translated scales in Malay now prevents populations who speak Malay from undergoing psychological testing (Kim et al. 2004). Second, the present study was able to address the lack of independent verification of the Subjective Happiness Scale's psychometric features outside of North America and Europe. Finally, the present study was interested in examining ethnic disparities in Subjective Happiness Scale answers within a particular cultural setting.

3.1.3 The Happiness Index Development

While employing a happiness meter is a way to affect societal change, the validity and dependability of the poll and the data gathered are equally crucial. The Happiness Alliance chose to offer this article as an explanation of the development processes underlying the Happiness Index after working in communities, with researchers, policymakers, and others. The objective is to educate index users on the accuracy and dependability of the tool and the data thus far gathered. This should serve as a reminder that the Happiness Alliance plans to do a more thorough statistical analysis in subsequent investigations.

3.1.4 Seventeen Sustainable Development Goals

In order to create a "blueprint to build a better and more sustainable future for all," the Sustainable Development Goals (SDGs), also known as the Global Goals, are a set of 17 interconnected international objectives. The UN General Assembly (UN-GA) established the SDGs in 2015, with the goal of achieving them by 2030. They are a part of the 2030 Agenda, often known as Agenda 2030, which is an UN-GA resolution. The SDGs were created as part of the Post-2015 Development Agenda to replace the Millennium Development Goals, which came to an end in 2015.

The 17 SDGs are No poverty, zero hunger, good health and well-being, quality education, gender equality, clean water and sanitation, affordable and clean energy, decent work and economic growth, industry, innovation and infrastructure, Reduced Inequality, Sustainable Cities and Communities, Responsible Consumption and Production, Climate Action, Life Below Water, Life On Land, Peace, Justice, and Strong Institutions, Partnerships for the Goals.

Fig. 3.2: Seventeen Sustainable Development Goals (Source- Wikipedia)

In order to evaluate and compare how successfully different cities, towns, neighbourhoods, and communities address sustainability challenges through community design and development, which may be related to residential happiness, the Sustainable Neighbourhoods for Happiness Index (SNHI) has been developed. Buildings, water, energy, and garbage are the systems with the most weight, according to the results of the restricted optimization, whereas urban planning and transportation have the least weight. The relative weightings for the SNHI show the areas on which planners, architects, engineers, and designers should concentrate their efforts to raise the SNHI score as a whole. Future communities as well as those evaluated in this study should be aware of these subsystems and work to enhance their SNHI. San Francisco has the highest SNHI, Detroit has the lowest (an extreme outlier), and Philadelphia, Sacramento, and Los Angeles are dispersed roughly around the mean. Plotted on the SNHD, Athens has a lower SNHI score than Ithaca based on the performance scores for both cities. The findings suggest that By concentrating on buildings & infrastructure, transportation, and business & economic development, Athens, Georgia, could enhance its total SNHI score.

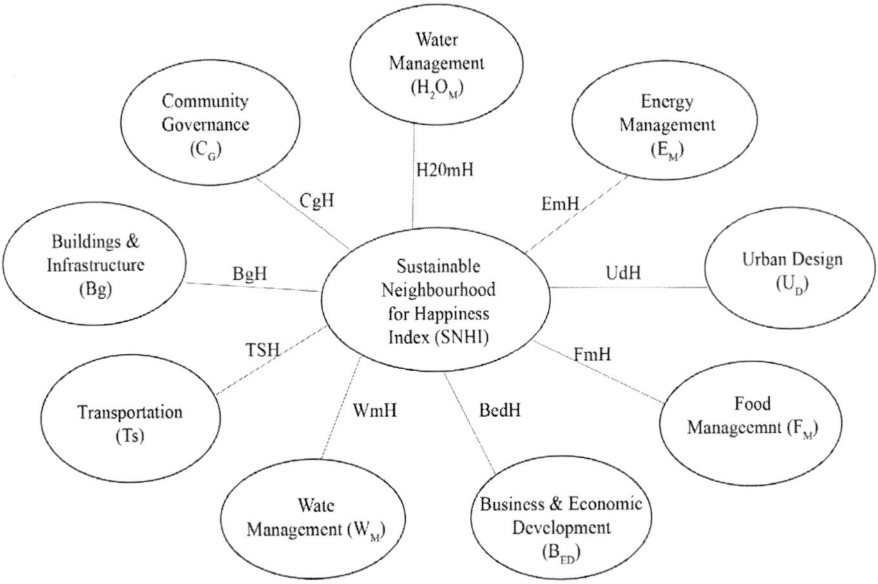

Fig 3.3: Sustainable Neighbourhood for Happiness Index (*Source*: Cloutier et al. 2014)

3.1.5 PERMA Theory of Well-Being

In an effort to provide answers, Dr. Seligman developed the PERMA theory of well-being. Positive emotion, engagement, relationships, meaning, and accomplishment are the five components that enable PERMA to flourish, and there are ways to raise each of these components. Each of these five fundamental components will contribute to the well-being of various persons to differing degrees. A successful life for one individual may not be successful for another. There are numerous paths to a prosperous existence. To put it another way, research on the elements that promote thriving can guide people in making decisions that are more in line with their beliefs and interests. In other words, we are not telling people what choices to make or what to value.

Fig. 3.4: PERMA Model (*Source*: Wikipedia)

Positive Emotion: This path to happiness is hedonistic—it fosters good emotion. Within reason, we can cultivate pleasant emotions about the past, the present, and the future. Examples include fostering gratitude and forgiveness, appreciating physical pleasures, and practicing mindfulness (e.g., by building hope and optimism).

Engagement: When someone totally commits their abilities, strengths, and attention to a difficult endeavour, they are engaged. When one's skills are just right for a hard activity, in pursuit of a distinct objective, with rapid feedback on progress toward the goal, flow is experienced. In such an activity, self-awareness vanishes, concentration is completely consumed in the present, and the perception of time is warped in hindsight, e.g., time stops. A nice conversation, a job activity, playing an instrument, reading a book, writing, creating furniture, fixing a bike, gardening, sports training, or performance are just a few examples of activities where the flow can be experienced.

Relationships: Relationships are essential to happiness. Our relationships frequently magnify the positive experiences that enhance our wellbeings, such as profound joy, significance, laughing, a sense of belonging, and pride in success. Relationships with other people can give life meaning and purpose. One of the finest ways to combat "the downs" of life and a dependable way to feel up is through support from and connection with people. According to research, showing kindness to others makes one feel better.

Meaning: Belonging to and serving a greater cause than oneself can give one a sense of meaning and purpose. A feeling of meaning can be enabled by a number of societal institutions, including religion, family, science, politics, work organizations, justice, the community, and social causes.

Accomplishment: People strive for success, competence, mastery, and achievement in a number of contexts, such as the office, sports, gaming, hobbies, etc. Even though achieving a goal does not inevitably result in great feelings, meaningful connections, or success, people nevertheless seek it.

3.2 History of World Happiness Index

Pierre le Roy, the man of GLOBECO, created the World Happiness Index. Since 2000, the world happiness index has been created to estimate overall happiness per country. It places a focus on overall happiness at the national level and seeks to measure development rather than the nation's statistical position. Gross National Happiness (GNH) is an index used to gauge a nation's overall happiness as well as a growth perspective. Bhutan is where the idea was first conceived. When Jigme Singye Wangchuck, the then-king of Bhutan, remarked during a conversation at the Bombay airport in 1979, the phrase "Gross National Happiness" was coined "I reject the concept of the gross national product. More significant is gross national happiness " The Bhutanese government has described the phrase as propaganda intended to draw attention away from the country's widespread violations of human rights.

3.3 India's Ranking on the Global Scale

Nations	Ranks
India	122
China	79
Pakistan	80
Nepal	99
Bangladesh	110
Iraq	117
Sri Lanka	120

On a list of the happiest nations in the world, India came in at a dismal 122. It now ranks fourth, four spots behind China, Pakistan, and Nepal, down from first last year. Norway is the happiest nation in the world, according to The World Happiness Report 2017. 155 nations are ranked according to their happiness levels in the 2017 World Happiness Report.

Norway rose up three ranks from the previous year, displacing Denmark. India was ranked 118 in the 2013–2015 study, however this year it dropped to the 122nd slot. Social support, GDP per capita, generosity, freedom to make life decisions, a healthy life expectancy, and views of corruption are among the factors that go into calculating happiness.

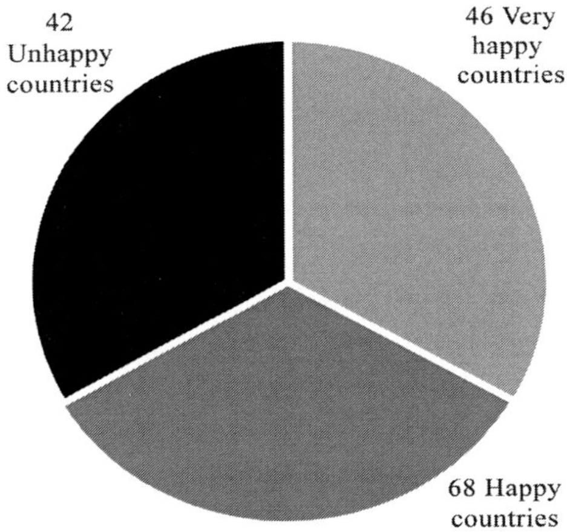

Fig. 3.5: Global Happiness Map (*Source*: Bhattacharyya et al. 2019)

3.4 Gross National Happiness (GNH)

Similar to the gross national product or gross domestic product, the Gross National Happiness (GNH) is an overall indicator of a nation's output. Gross national happiness (GNH) aims to quantify the totality of factors including economic output, net environmental effects, the spiritual and cultural development of the populace, physical and mental health, as well as the resilience of the economic and political systems. The UN General Assembly issued a resolution in July 2011 requesting that member nations assess the level of happiness among their citizens and utilize that data to inform public policy. The first UN High-Level Meeting on "Happiness and Well-Being: Defining a New Economic Paradigm" was held on April 2, 2012, and it was presided over by Bhutan's Prime Minister Jigme Thinley. Bhutan is the first and only nation to have formally adopted GNP (Gross National Product) rather than GDP (Gross Domestic Product) as its primary development indicator. On April 1, 2012, the first World Happiness Report was made public. Although an index to gauge happiness at work may seem trite, the experiences of some of the most desirable employers in the world demonstrate that businesses can no longer disregard the happiness quotient in their workplaces.

A GNH policy's implementation can be difficult because it needs a lot of institutional backing. The mainstreaming of GNH into political institutions in Bhutan has been a long-term, methodical process that has lately accelerated with the launch of the GNH Index and the GNH Screening Tool.

The Centre for Bhutan Studies (CBS) created the GNH Index as part of a protracted and ongoing process to incorporate the GNH ideology into public policy in order to gauge the development of Bhutanese society. The first nationwide GNH survey was carried out in 2010 with a sample size of 8,510 Bhutanese adults. The sample size for the second national survey, which was performed in 2015, was 8,871.

Following the completion of all three rounds of surveys, extra data collection, including follow-up interviews, was done in order to review and improve the survey. Results for many demographic parameters, including gender, age, place of residence, and occupations are given for the GNH survey, which is conducted in all twenty districts (Dzonkhag). The poll thus offers a comprehensive dataset to analyze the level of happiness among various citizenry groups and how it has evolved over time.

3.4.1 Measuring Gross National Happiness Index

33 indicators are used to measure the Gross National Happiness Index, and they come from nine different key categories. The components of happiness are the domains within GNH, and the index equally weights each one.

The nine domains of GNH are:

a. Education
b. Good governance
c. Cultural diversity and resilience
d. Psychological well-being
e. Time use
f. Health
g. Community vitality
h. Living standard
i. Ecological diversity and resilience

3.4.2 Criticisms of Gross National Happiness Index

The Gross National Happiness Index is highly regarded in Bhutan, but it has drawn harsh criticism from other countries. The fact that the domains and indicators are largely subjective is one of the main criticisms of GNH. The subjective nature of the indicators, according to critics, makes it much more challenging to obtain a precise quantitative evaluation of happiness. Additionally, they assert that because of the subjectivity, governments may be able to alter GNH results to their advantage.

Other critics contend that using Bhutan's measures as benchmarks to measure happiness and development in other nations is problematic because the definition and, consequently, ranking of happiness vary per country. People in Bhutan or India may rank education or living standards differently than those in France. However, despite these criticisms, it is crucial to remember that GNH is a modified and significant way to simply look at economic and social progress throughout the world.

3.5 Employee Happiness

The majority of businesses now recognize how crucial employee satisfaction is. Most seminar speakers point out that being happy can improve your health, make you kinder and more productive, and even increase your chances of getting promoted. Organizations spend money on Chief Happiness Officers, happiness coaches, team-building activities, and games.

When you carefully examine the studies, there is evidence to show that contented workers are safer, more likely to satisfy consumers, and less likely to leave. They are also more inclined to engage in civic behavior. However, we also found contradictory results that suggest some of the conventional wisdom about what workplace happiness may accomplish are merely falsehoods.

3.5.1 Inexpensive ways to keep employees happy

- Trust people – give them freedom within guidelines
- Help people see why what they do matters
- Give regular encouragement, praise, perks, and thanks
- Help people find and play to their strengths
- Encourage a healthy balance between work and life
- Treat your employees with respect-Give respect and take respect, treat them well.
- Money Matters-provide adequate vitamin M to all employees as per their performance.
- Build ownership among your employees- let employees feel that they own the place, not just work there.
- A positive work environment has long-lasting effects on employee engagement, overall productivity and profit &ultimately happiness.

3.6 Measuring Happiness

One in five Indians who require mental health counseling have a mental condition that affects less than 5% of them. In India, mental illness will cost the country 38.1 million years of healthy life by 2025. Due to the rising stress of modern life, mental serenity and happiness are now necessities. If we want to stop the ticking time bomb of mental illness from erupting, we need to understand happiness, measure it, and find ways to be happy in our lives. Happiness is characterized by positive or pleasurable emotions, ranging from contentment to great delight, according to psychologists. Happier humans have better physiological functioning, are better able to handle stress and discomfort, and have longer life spans.

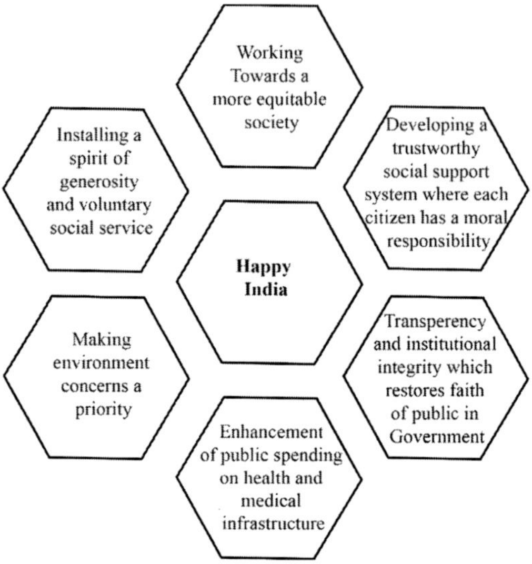

Fig. 3.6: Six-pronged strategy for India to go the Nordic way (Bhattacharyya et al., 2019)

3.6.1 Efforts in Measuring Happiness

According to sociologists, happiness is "the degree to which a person positively evaluates the overall quality of his life-as-a-whole, or how well one likes one's life." It is frequently referred to as life satisfaction or subjective well-being (SWB). There aren't many scales available to measure the SWB of people in various nations. The Subjective Happiness Measure (SHS) is a 4-item scale that asks respondents to categorize themselves as either happy or dissatisfied using absolute scores (out of 7) for each question. The Positive and Negative Affect Schedule (PANAS) is a 20-item questionnaire that uses a five-point Likert scale to assess the relationship between personality factors and positive or negative effects. In order to complete the Satisfaction With Life Scale (SWLS), a person must indicate whether they agree or disagree with five items about their life. A few surveys also keep tabs on people's overall contentment with their lives. The World Values Survey (WVS) is an international research initiative that examines people's values and beliefs, how they evolve over time and the effects they have on society and politics. The same is carried out by Euro barometer in the EU Member States.

According to research, achieving sustainability does not necessitate sacrificing happiness, and design tactics can simultaneously increase both (Zidansek, 2007). Additionally, studies by Leyden et al. (2011), Florida (2010), O'Brien (2001), Schimmel (2009), and Zidansek (2007) revealed links between happiness levels and sustainable development. Bhutan and Thailand are noteworthy examples of nations that have developed and applied happiness indices. Bhutan measures nine factors that are related to happiness for determining its Gross National Happiness: (1) Mental health or psychological well-being; (2) Physical health; (3) Time and work-life balance; (5) Education; (6) Social connection and relationships; (7) Environmental quality and access to nature; (8) Governmental quality; and (9) Material well-being (DeGraaf *et al.*, 2005).

3.7 Farmers Happiness

By 2030, there will likely be 5 billion urban dwellers worldwide, up from 2.6 billion in 2000. (Seto et al., 2010). Metropolitan land is in high demand due to the urban population's quick increase, which has led to the city's ongoing development (Zhao et al., 2017). Additionally, the area around the city is encroached upon by urban expansion (Wu et al., 2016, p.1). It is anticipated that developing nations in Asia and Africa would experience the majority of the world's urban population growth, urban expansion, and urban development (Van Ginkel, 2008; Seto et al., 2012). Due to their vast populations, rising rates of population urbanization, and rapid urban growth, China and India take the lead in urban development in Asia (Zhang et al., 2020). India's urban population currently makes up less than one-third of the country's overall population, but by 2050 that number will rise to more than half (Mitra et al., 2016). People from rural areas go to cities in search of better employment prospects, greater incomes, and better quality of life. It increases the amount of agricultural land needed for the construction of residential, commercial, and industrial projects (Chen et al., 2014). As a result, India's agricultural land conversion and uses are under immense strain as a result of urban expansion and growth (Fazal, 2000; Chadchan and Shankar, 2012). The main elements influencing urban land expansion include population increase, economic development, rise in income, land-use policies, capital flow, and transportation expenses (Seto et al., 2011). By 2030, croplands will have lost between 1.8 and 2.4 percent of their area due to rapid urbanization. By 2030, emerging nations in Asia and Africa will account for almost 80% of all cropland lost due to urbanization (Bren d'Amour et al., 2017). Land acquisition in developing nations like India is driven by the construction of highways, roads, airports, ports, markets, industrial parks, hospitals, schools, public parks, police stations, waste management sites, and community centers, none of which are likely to be provided by the private sector. The establishment of recognized special economic zones (SEZs) in various states, land purchase and conversion for urban expansion, and a deterioration in the appropriateness for agricultural activities are the main causes of land loss in India (Fazal, 2000; Panday and Seto, 2014). (Prokop and Poremba 2012) A few factors contributing to the abandonment of farming include rising cultivation costs, recurring crop failure, productivity risks, and farmers' increasing debt loads (Parwez, 2015). In India's rural areas, the proportion of cultivators to all workers fell from 40.2 percent in 2001 to 33 percent in 2011.

Antidumping duties are becoming a more popular choice among import-competing businesses in the US as a "simple" way to obtain protection (Irwin, 2005). The Continued Dumping and Subsidy Offset Act, passed by the US Congress in October 2000, is one of the causes, though there are many others (Zanardi 2004; Prusa 2005). The "Byrd Amendment," or BA, allows producers and processors who successfully petition for antidumping levies to receive tariff revenues. An additional motivation to apply for protection and take part in the petition process is the ability to earn tariff receipts (Reynolds 2006).

According to Schmitz and Seale (2004), payments made under the BA between 2001 and 2003 totaled $749 million, $299 million of which went to two bearing manufacturers combined in 2003. The satisfaction that farmers derive from fish farming is not significantly impacted by either revenue per capita or absolute income from fish farming. However, the model results reveal that relative income from fish farming increases the overall likelihood of a farmer being satisfied with their work, proving that relative income—and not absolute money—is related to job happiness. Farmers who participated in AOP's aquaculture extension activities, who had higher income expectations from fish farming, and who have larger pond surfaces compared to total land area all show higher levels of pleasure. On average, older farmers are happier and more satisfied with their fish farming practices.

3.8 Farmer's Agony

Suicide among female farmers is on the rise. Studies of women in farming have found high levels of stress, fatigue, and depression. Explanations most commonly given are role conflicts and high workload. Farm women, unlike men, experience stress not only due to the farm operations but also due to the impact of farming stressors on the physical, social and financial wellbeing of all family members. As farming has become less profitable, women are taking more and more on and off-farm work to supplement the family income. Farming women in this position often become stressed and fatigued due to multiple tasks and the conflict between their traditional role as housemaker and the need for off-farm income. The female farmers have an additional burden of performing household chores apart from farming. This makes farm women a high-risk category for suicide. England, Australia, and now India had reported suicide among this gender. Farming is a high-stressed and dangerous occupation.

4

Social Ecology

4.1 Research Setting

In any social science research, it is hardly possible to conceptualize and perceive the data and interpret the data more accurately until and unless a clear understanding of the characteristics in the area and attitude or the behavior of people is at commending by the interpreter who intends to unveil an understanding of the implication and behavioral complexes of the individual who lives in the area under reference and form a representative part of the larger community. The socio-demographic background of the local people in a rural setting has been critically administered in this section. A research setting is a surrounding in which input and elements of research are contextually imbibed, interactive, and mutually contributive to the system performance. The research setting is immensely important in the sense because it is characterizing and influencing the interplays of different factors and components.

4.2 Area of Study

The present study was taken up in the Nadia district under the alluvial zone of West Bengal. The study was carried out in Haringhata blocks from Nadia. A brief description of the whole study area has been provided in this chapter.

4.3 Profile of the State, West Bengal

The state of West Bengal was formed on 26th January 1951. Kolkata (formerly known as Calcutta), popularly known as 'the city of joy' which once served as the capital of the country in the British period, still holds its glory as the capital of the state. Presently, the state has been bifurcated into 23 districts under five divisions – Burdwan division, Jalpaiguri division, Presidency division, Medinipur division, and Malda division. The districts are – Alipurduar, Bankura, Birbhum, Cooch Behar, Darjeeling, Dakshin Dinajpur, Hooghly, Howrah, Jalpaiguri, Jhargram, Kolkata, Kalimpong, Malda, Murshidabad, Nadia, North 24 Parganas, PaschimBardhaman, Paschim Medinipur, PurbaBardhaman, Purba Medinipur, Purulia, South 24 Parganas, and Uttar Dinajpur. The state stretches over an area of 88,752 sq. km. with 66 sub-divisions, 121 municipalities, 6 municipal corporations, 341 development blocks, 40, 782

Mouzas, 3349 Gram Panchayats, and 37, 469 inhabited villages. West Bengal is on the eastern bottleneck of the country, stretching from the Himalayas in the north to Bay of Bengal in the south. The state extends between $86^0 00'$ and $89^0 83'$ Eastern longitude and $21^0 86'$ and $27^0 13'$ North latitude. It shares its northern boundary with Sikkim and parts of Nepal and Bhutan, eastern boundary with Bangladesh, western boundary with states of Orissa, Bihar and Jharkhand and to the southern part of the state lies the Bay of Bengal. (www.wb.gov.in)

Political Map of West Bengal

4.3.1 Demographic Features

As per the census of 2011, the total population of the state is 91, 276, 115 of which 46, 809, 027 and 44, 467, 088 are male and female respectively. It has a share of 7.54% of the total country's population with a population density of 1028 per sq. km. 68. 13% of the total population of the state counts for its rural population. The sex ratio of the state is 950 per thousand. Per capita income at the current price is Rs. 61,352.35 as of August 2014. The literacy rate of the state is 76.26 percent, which is higher than the national average. The percentages of total literate males are 81.69 percent and that of females is 70.54 percent. (www.wb.gov.in)

4.3.2 Climate of West Bengal

The climate of the state comes under two distinct types. The climate of the northern part of the state is characterized by the sub-tropical monsoon, mild to dry winter, and hot summer; whereas the southern or the Gangetic part of the state experiences a tropical savanna with hot and dry winter. However, the northern most part viz., Darjeeling district has a climatic type of tropical upland, mild and dry winter, and short warm summer. The year may be divided into four seasons. The winter season stretches from December to February, followed by the pre-monsoon season from March to May. The period from June to the middle of September constitutes the southwest monsoon and the period from the latter half of September to November is the post-monsoon period. The period from December to February is generally very unpleasant due to low temperatures over the sea except in the coastal belt. In the hot weather season from March to May, the weather is dry and uncomfortable in the interior. Due to lower temperatures, the hilly regions are, however, comparatively less uncomfortable. The weather tends to be oppressive during June due to high humidity and temperature. The rest period of the monsoon is fairly comfortable due to reduced day temperatures, although humidity continues to be high. Day temperatures are more or less uniform over the plains during the monsoon and post-monsoon seasons (increases westwards in pre-monsoon and southwards in winter). In the state, the night minimum temperatures are lower in higher latitudes. Both day and night temperatures are lower at high-level stations than over the plains. May and April are the hottest months with the mean maximum temperatures of 35.6°C and 33.8°C in the Gangetic and sub-Himalayan West Bengal respectively. The relative humidity is generally high during the period from July to September. It is about 80% in June rising to about 83%-85% in July, August, and September in the morning. The diurnal variation of relative humidity is least during the monsoon season. The relative humidity is lowest during the summer afternoons when it becomes 6 about 38 to 50% at the plain.

4.3.3 Agriculture Scenario of the State

West Bengal is an agrarian state. More than about 72% of the rural population of the state still earns their bread from agriculture. The total cultivable area in the state contributes to 65.65% of the total area. The net sown area of the state is 5.234mha. Although the topography of West Bengal varies hugely, blessed by the river Ganga, fertile alluvium predominates over a large area making the soil of the state highly productive. Moreover, due to prevailing both tropical and temperate climatic conditions, diversification in agriculture is another important feature of the state. Seeing its high potential in the agriculture sector, the central government has included the state as one of the seven states in the "Bringing Green Revolution to Eastern India (BGREI)" Programme of 2011.

Table 1: Agricultural statistics of West Bengal:

1.	Total reported area	8.68 mha
2.	Total cultivable area	5.65 mha
3.	Total forest area	1.17 mha
4.	Cultivable wasteland	16940 ha
5.	Permanent pasture and other grazing land	2370 ha
6.	Gross cultivated area	9.6 mha
7.	Net cultivated area	5.2 mha

Statistical Abstract, 2015

The average cropping intensity of the state is 185% which is much higher than the national average (142%). The fertilizer consumption ratio (N, P, K ratio) of the state is 2.3:1.22:1 which is much lower than the recommended dose of 4:2:1, which clearly indicates the imbalance in nutrient supply to the plants, lowering the production and productivity. This discrepancy in fertilizer consumption also shows the nutrient deficiency in the soils of West Bengal. Besides this intensive cultivation, nutrient imbalance, and nutrient deficiencies, the reliance of the farmers on chemical fertilizers has increased the acidity of the soil in many parts of the state. In addition to this, in the southern part of the state, the *Sunderbans*region faces problems of soil salinity because of being exposed to the brackish water of the Bay of Bengal. The major crops of the state include food crops like paddy, wheat, maize, pulses like lentil, *mung,* arhar, gram, Lathyrus etc., potato, vegetables viz. brinjal, tomato, okra, cabbage, cauliflower, cucurbits, onion, etc. and other leafy vegetables and oilseeds viz. rapeseed and mustard, sesame, linseed and others. Jute is the most important cash crop of the state. The state is also the house of the world-famous Darjeeling tea. Hooghly and Burdwan districts of the state are considered to be the most productive region which also forms the 'Rice Bowl' for the state. The other agriculturally rich districts are Nadia, Medinipore, both 24 Paraganas, Coochbehar, etc. The state is also rich in horticulture, producing mango, banana, pineapple, papaya, guava, jackfruit, litchi, citrus, sapota, etc. The state also houses different livestock like cattle, buffaloes, sheep, pigs, and goats, in a considerable number.

Description of Agro-ecological Zones

West Bengal is divided into 6 agro-ecological zones with unique and distinct features. The 6 agro-ecological zones are – the Hill zone, Terai zone, Old alluvial zone, New alluvial zone, Red and lateritic zone, and Coastal and saline zone. The present study was conducted in New alluvial zone and Terai zone.

The New alluvial zone comprises the districts of Nadia, Howrah, Hooghly, North 24 Parganas, Kolkata, parts of Bardhaman, and parts of South 24 Parganas. The zones stretches over an area of 15.304 lakh ha, which is 55.7% of the total geographical area of the state. The area experiences an annual precipitation of 1350-1450mm. The average temperature ranges between 35 degree Celsius to 15 degree Celsius. The soil texture is loamy to clay in nature. The region is blessed with the fertile alluvium of the Ganges, and grows a variety of agricultural crops from rice, wheat, mustard, potato, and seasonal vegetables as well as cash crops like jute and Mesta.

The terai region comprises the districts of Jalpaiguri, Coochbehar, Alipurduar, and Uttar Dinajpur. The zone stretches over an area of 0.17 million hectares, covering 1.9% of the total geographical area of the state. The annual precipitation in this region ranges between 2000-3200mm of rainfall and the temperature varies between 30 degrees Celsius and 10 degrees Celsius or below. The soil texture of this region is mainly sandy to sandy loam in nature. The region is gradually becoming agriculturally rich by adopting different sustainable agricultural practices like conservation agriculture.

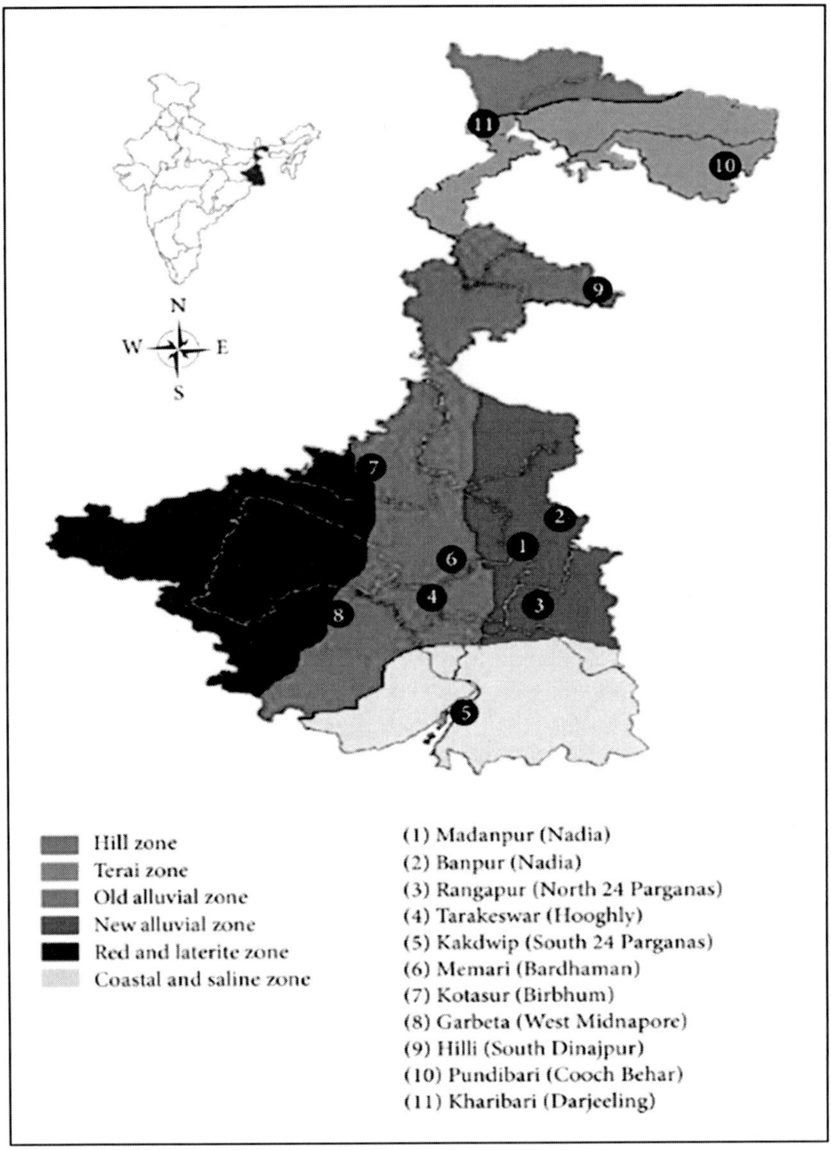

Agro-ecological Map of West Bengal

4.4.1 Description of Nadia District

The district of **Nadia** is situated in the heart of the Bengal delta and held within the arms of the Ganga, namely, the Bhagirathi on the West and Mathabhanga on the North. The entire district lies in the alluvial plain of the Ganga and its tributaries. Nadia district takes its name from the memory of Lord Shri Chaitanya Mahaprabhu who was born here on 18th February 1486. The British district of Nadia was formed in 1787. The present district of Nadia after partition was formed on 23rd February 1948. The district of Nadia is bounded on the north and north-west by the district of Murshidabad. On the south-east and east, it is bounded by the Republic of Bangladesh. In the south and south-east, the district is bounded by the district of North 24-Parganas. The shape of the district is irregular, lying North to South. The district is about 46ft. above the mean sea level and the tropic of cancer divides the district into two parts.

4.4.2 Demography of Nadia district

The district has an area of 3927 and seems to have a population of 46,04,827 as per Census 2001. Out of that SC & ST population are 13,65,985 and 1,13891 respectively. The district has 17 PanchayetSamities consist of 187 Gram Panchyets and 8 Municipalities. Total number of Police Stations in the district is 19. The density of population in this district is 1173 persons per sq km. Nadia district has 950 females as against 1000 males. The majority of the people of the district speaks Bengali followed by Hindi, Santali and other. About 73.75% are Hindus and 25.42% are Muslims. In the district of Nadia the percentage of literacy by sex is 72.30 (Male) & 59.60 (Female) as per Cencus 2001. The important rivers of the district are Bhagirathi, Churni,

Mathabhanga, Ichamati and jalangi.

Table 4.4.3: General Information of Nadia District

Total population	5,167,600
Male	2,653,768
Female	2,513,832
Population Growth	12.22%
Sex ratio(per 1000)	947
Population density	1300/ km^2
Literates Rate	74.97%
Male	78.75%
Female	70.98%
No of Sub divisions	5
No of Development blocks	17
No of Gram Panchayats	187
Inhabited Villages	2639
Area (Sq. Km)	3,927

Source: Office of the Register General and Census Commissioner 201, Ministry of Home Affairs, GOI

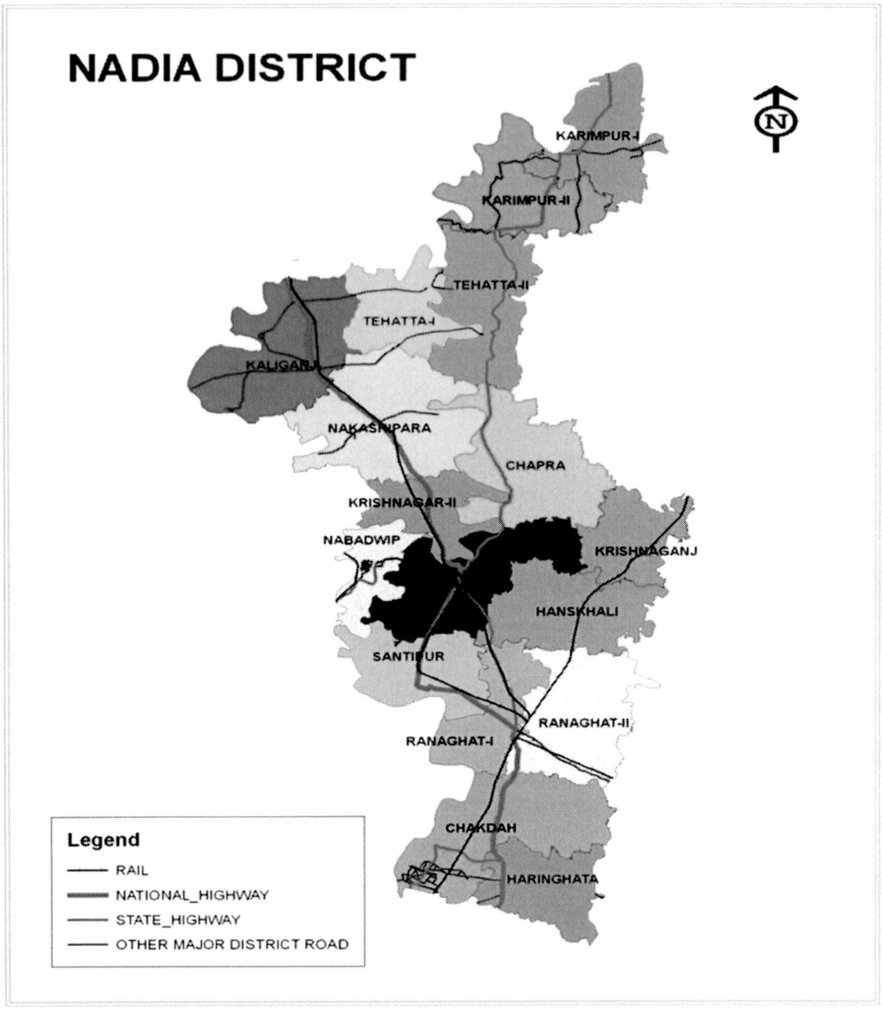

Map of Nadia District

4.4.4 Climate of Nadia District

The climate of this district is characterized by an oppressively hot summer, high humidity nearly all the year round, and a well-distributed rainfall during the southwest monsoon season. The winter season is from about the middle of November and continues till the end of February. The period from March to May is the summer season. The southwest monsoon season commences by about the beginning of June and lasts till the end of September. October and the first half of November constitute the post-monsoon season. The average annual rainfall in the district is 1369.4 mm. Haringhata in the southern part of the district gets 1698.6 mm on average in a year while Ranaghat further north in the middle portion of the district has an annual rainfall of 1161.0 mm. The rainfall during the southwest monsoon season, June to September constitutes about 76% of the annual rainfall, with July being the rainiest month. The variation in the rainfall from year to year is large. The cold season commences by about the middle of November when the temperature begins to decrease. January is the coldest month with the mean daily maximum temperature at 28.2°C and the mean daily minimum temperature at 10.7°C. In association with the passing of western disturbances in the cold season, the district is sometimes affected by cold waves and on such occasions, the minimum temperature may go down up to about 1 to 3°C. By about the end of February, the temperatures begin to rise. While the night temperatures reach a maximum only in the southwest monsoon season, day temperatures usually reach the maximum in April when the mean daily maximum temperature is 38.0°C. The heat in summer is often oppressive on account of the high moisture content of the air. There is a welcome relief from the heat though temporarily when thundershowers occur on some days in this season. With the onset of the southwest monsoon by about the first week of June, day temperatures begin to drop but night temperatures continue to rise. With the increased humidity in the air and the continuing high night temperatures, even during the monsoon season, the weather is often uncomfortable in between the rains. The southwest monsoon withdraws early in October and the temperatures begin to drop. The drop particularly in the night temperatures is more rapid from about the middle of November. The values of relative humidity are generally high throughout the year, particularly in the mornings, and are about 65% to 80%. But in the summer months, March and April the values of relative humidity are comparatively less, particularly in the afternoons being about 40 to 45%.

4.5 Profile of Haringhata Block

Haringhata is a town in the Kalyani subdivision of Nadia district in the state of west Bengal of India.

Location

Haringhata is located at 22°95′N 88°57′E. It has an average elevation of 10 meters. Haringhata community development block has an area of 168.59 Km^2.

Language

Bengali is the local language in these areas.

Table 4.5.1: General Information about Haringhata block

Block area	170.32 sq. Km
No. of gram panchayats	10
No. of gram sabhas	129
No. of Mouza No. of police stations	87 1
Elevation	13 m (43 ft.)
Total Population (2011)	231,068
Density	1,400/km^2(3,500/sq. mi)
Languages	Bengali, English
Total literates	171,144 (82.15%)

4.5.2. Geography of Haringhata Block

Haringhata Block is bounded by Chakdah Block in the north and west, Bangaon and Gaighata Blocks, in North 24 Parganas district, in the east, Amdanga, Habra I and Habra II Blocks, in North 24 Parganas district, in the south. Nadia district is mostly alluvial plains lying to the east of Hooghly River, locally known as Bhagirathi. The alluvial plains are cut across by such distributaries as Jalangi, Churni, and Ichhamati. With these rivers getting silted up, floods are a recurring feature. Haringhata Block has an area of 170.32 km^2. It has 1 panchayat samity, 10 gram panchayats, 176 gram sansads (village councils), 87 mouzas and 82 inhabited villages. police station serves this block. Headquarters of this Block is at Subarnapur. It is located 56 km from Krishnanagar, the district headquarters.

Gram panchayats of Haringhata Block/ Panchayat Samiti are: Birohi 1, Birohi 2, Fatepur, Haringhata 1, Harinhgata 2, Kastodanga 1, Kastodanga2, Mollabelia, Nagarukhra 1 and Nagarukhra 2.

4.5.3. Demographics

i Population

As per the 2011 Census of India, Haringhata Block had a total population of

231,068, of which 207,459 were rural and 23,609 were urban. There were 118,709 (51% males and 112,359 (49%) females. Scheduled Castes numbered 73,755 (31.92%) and Scheduled Tribes numbered 11,218 (4.85%).

As per the 2001 census, the Haringhata block had a total population of 207,867, out of which 108,011 were males and 99,856 were females. Haringhata block registered a population growth of 16.92 percent during the 1991-2001 decade. Decadal growth for the district was 19.51 percent. Decadal growth in West Bengal was 17.84 percent.

There are four census towns in Haringhata Block (2011 census figures in brackets): Haringhata (3,989), Simhat (6,949), Subarnapur (5,759), and Digha (6,916).

Large villages (with 4,000+ populations) in Haringhata Block were (2011 census figures in brackets): Chandirampur (4,740), Narayanpur (4,548), Uttar Datta Para (5,430), Uttar Brahmapur (5,371), Subuddipur (6,051), Sekendarpur (7,917), Kathdanga (7,417), Mollabalia (5,889), Ganguria (4,291), Fatepur (5,093), Barasat (4,046),Ukrah (13,548), Goaldob (4,883) and Dighalgram (5,390).

ii Literacy

As per the 2011 census, the total number of literates in the Haringhata CD Block was 171,144 (82.15% of the population over 6 years) out of which males numbered 93,313 (87.11% of the male population over 6 years) and females numbered 77,831 (76.90% of the female population over 6 years). The gender disparity (the difference between female and male literacy rates) was 10.21%.

iii. Language and Religion

Bengali is the local language in these areas. In the 2011 census, Hindus numbered 164,263 and formed 71.09% of the population in Haringhata Block. Muslims numbered 65,148 and formed 28.19% of the population. Christians numbered 460 and formed 0.20% of the population. Others numbered 1,197 and formed 0.52% of the population.

In the 2001 census, Hindus numbered 151,379 and formed 72.79% of the population of Haringhata CD Block. Muslims numbered 55,885 and formed 26.97% of the population. In the 1991 census, Hindus numbered 130,125 and formed 73.19% of the population of Haringhata CD Block. Muslims numbered 47,284 and formed 26.60% of the population.

4.5.4. Rural Poverty

The District Human Development Report for Nadia has provided a CD Block wise data table for the Modified Human Vulnerability Index of the district. Haringhata CD Block registered 31.59 on the MHPI scale. The CD Block wise mean MHVI was estimated at 33.92. A total of 8 out of the 17 CD Blocks in Nadia district were found to be severely deprived when measured against the CD Block mean MHVI Karimpur I and Karimpur II (under Tehatta subdivision), Kaliganj, Nakashipara, Chapra, Krishnanagar I, and Nabadwip (under KrishnanagarSadar subdivision) and Santipur (under Ranaghat subdivision) appear to be backward.

4.5.5. Economy

i) Livelihood

In Haringhata CD Block in 2011, amongst the class of total workers, cultivators formed 18.42%, agricultural laborers 35.60, household industry workers 3.79%, and other workers 42.19%.

The southern part of Nadia district starting from Krishnanagar I down to Chakdah and Haringhata has some urban pockets specializing in either manufacturing or service-related economic activity and has reflected a comparatively higher concentration of population but the urban population has generally stagnated. Nadia district still has a large chunk of people living in the rural areas.

ii) Infrastructure

There are 82 inhabited villages in Haringhata Block. 100% of villages have power supply and drinking water supply. 13 Villages (15.85%) have post offices. All 82 villages (100%) have telephones (including landlines, public call offices, and mobile phones). 64 villages (78.05%) have a pucca approach road and 30 villages (38.59%) have transport communication (includes bus service, rail facility, and navigable waterways). 6 villages (7.32%) have agricultural credit societies and 8 villages (9.76%) have banks. It should, however, be noted that although 100% of villages in Nadia district had power supply in 2011, a survey in 2007 08 revealed that less than 50% of households had electricity connection. In rural areas of the country, the tube well was for many years considered to be the provider of safe drinking water, but with arsenic contamination of groundwater claiming public attention it is no longer so. Piped water supply is still a distant dream. In 2007 08, the availability of piped drinking water in Nadia district was as low as 8.6%, well below the state average of around 20%.

iii) Agriculture

Although the Bargadari Act of 1950 recognized the rights of bargadars to a higher share of crops from the land that they tilled, it was not implemented fully. Large tracts, beyond the prescribed limit of land ceiling, remained with the rich landlords. From 1977 onwards major land reforms took place in West Bengal. Land in excess of the land ceiling was acquired and distributed amongst the peasants. Following land reforms, land ownership pattern has undergone a transformation. In 2013 14, persons engaged in agriculture in Haringhata Block could be classified as follows: bargadars 2.44%, Patta (document) holders 7.95%, small farmers (possessing land between 1 and 2 hectares) 6.18%, marginal farmers (possessing land up to 1 hectare) 33.53%

and agricultural labors 49.09%. As the proportion of agricultural labor is very high, the real wage in the agricultural sector has been a matter of concern.

Haringhata CD Block had 207 fertilizer depots, 40 seed stores, and 44 fair price shops in 2013-14.

In 2013-14, Haringhata CD Block produced 72,866 tonnes of Aman paddy, the main winter crop from 24,942 hectares, 20,348 tonnes of Boro paddy (spring crop) from 5,720 hectares, 5,413 tonnes of Aus paddy (summer crop) from 2,013 hectares, 4,317 tonnes of wheat from 1,315 hectares, 42,799 tonnes of jute from 2,383 hectares and 16,616 tonnes of potatoes from 536 hectares. It also produced pulses and oilseeds.

In 2013-14, the total area irrigated in Haringhata Block was 967 hectares, out of which 89 hectares were irrigated by river lift irrigation, 830 hectares by deep tube wells, and 48 hectares by shallow tube wells.

4.5.6. Haringhata Dairy Farm

In order to supply milk to the urban population in West Bengal, the first dairy plant was installed in 1950 at Haringhata by the Directorate of Dairy Development, under the aegis of the Animal Resources Development Department, Government of West Bengal. It started with a production of 1.5 thousand liters of bottled milk per day in 1950 51. With a network of two dairy plants, seven milk colony units, and 19 Milk Collection cum Chilling stations named —Greater Calcutta Milk Supply Scheme‖ the production had gradually gone up to 210 thousand liters per day by the 1980s. (Directorate of Dairy Development, under the aegis of Animal Resources Development Department, Govt. of West Bengal)

4.5.7. Banking

In 2013-14, Haringhata Block had offices of 12 commercial banks and 2 Gramin banks.

4.5.8. Transport

Haringhata CD Block has 4 originating/ terminating bus services. The nearest railway station Kanchrapara is 16 km and Kalyani Railway Station is 19 km from Block headquarters.

NH 12 (old number NH-34) passes through this block. SH1, running from Bangaon (in North 24 Parganas district) to Kulpi (in South 24 Parganas district) passes through this Block.

4.5.9. Education

In 2013 14, Haringhata Block had 120 primary schools with 9,251 students, 11 middle schools with 866 students, 3 high schools with 1,819 students, and 18 higher secondary schools with 23,167 students. Haringhata Block had 1 general college with 3,233 students, 2 technical/ professional institutions with 1,880 students, and 453 institutions for special and no normal education with 10,874 students. In Haringhata Block, amongst the 82 inhabited villages, 4 villages did not have any school, 21 had more than 1 primary school, 16 had at least 1 primary and one middle school and 14 had 1 middle school and 1 secondary school.

4.5.10. Healthcare

In 2014, Haringhata Block had 1 rural hospital, 4 primary health centers, and 2 private nursing homes with a total of 96 beds and 9 doctors (excluding private bodies). It had 30 family welfare sub-centers. 4,907 patients were treated indoors and 357,557 patients were treated outdoor in the hospitals, health centers, and sub-centers of the Block. Haringhata Rural Hospital at Mohanpur has 25 beds. There are primary health centers at Birohi (6 beds), Kasthodanga (6 beds), and Nagarukhra (10 beds). Haringhata Block is one of the areas of Nadia district where groundwater is affected by the high level of arsenic contamination. The WHO guideline for arsenic in drinking water is 10 mg/liter, and the Indian Standard value is 50 mg/liter. All the 17 blocks of Nadia district have arsenic contamination above this level. The maximum concentration in Haringhata Block is 865 mg/liter.

4.6. Profile of Village

4.6.1. Dalilpur Village

The village Dalilpur falls under Haringhata Municipal Corporation, Ward No.-07, Haringhata block of Nadia district. The PIN code of Dalilpur is 741246. The total geographical area of this village is approximately 600 bigha. There are about 560 houses in Dalilpur village. Gayeshpur is the nearest town to Dalilpur which is approximately 3 km away. The total population of the village is 2000 approximately, where the male population is about 1100 and the female population is about 900. Public bus service is available within 3 km distance and railway station within 8 km distance. There are 3 schools in the village.

Table 4.6.2: Cohort study of Dalilpur village

COHORT				
SL No.	Particulars/ Indicators	Years		
		1990 - 2000	2000 - 2010	2010 - 2020
1	Crop Combination	Boro rice (cultivated by 10% of farmers), jute, parwal, cauliflower, bitter gourd, brinjal, mustard,	Boro rice (cultivated by 15% of farmers), banana, yam, cabbage, cauliflower, jute	Boro rice (cultivated by 90% of farmers), guava (cultivated by 70% of farmers).
2	Rice Yield (Per Bigha)	16 bostha 1 bostha = 60 kg	14 bostha	10-12 bostha
3	Cost of Cultivation of Rice (Rs./bigha)	6000	8000	12000
4	Soil Health	Very good	good	poor
5	Water Availability (no. of shallow) tube well	50	80	150
6	Tree and crop biodiversity	Tree: Shishu, Shirish, Akashmoni Rice variety: Murkibalam, Jaya, Khitish	Tree: Babla, Eucalyptus, Mahogany Rice variety: Minikit, Ranjit, Pankaj	Tree: Lombu, Kadam Rice variety: Satabdi, Gosai
7	Application of Organic Manure (cow dung)	More	Medium	Very less
8	Livestock Count (cows)	800	400	60
9	No. of ponds	2	3	5
10	Rainfall (Rainy Days)	High (4-5 months)	Medium (3-4 months)	Less (2 months)
11	Income/Net Return (Rs./month)	4000	6000	8000
12	Expenditure (Rs./month)	1500	4000	6000
13	Well-being (Score) (out of 10)	5	7	9

Source: Unush Ali Mondal and Muazzam Hossain, farmers of Dalilpur Village

4.6.3 Kastodanga Village

Kastodanga is a village in Haringhata Block in Nadia District of West Bengal State, India. The village Kastodanga falls under the gram panchayat Kastodanga-I. It belongs to Presidency Division. It is located 59 KM South of District head quarters Krishnanagar, Nadia. 7 KM from Haringhata.60 KM from the State capital Kolkata. Kastodanga-I Pin code is 741257 and the postal head office is Nagarukhra. Kastodanga-Iis surrounded by Habra-I Block towards South, Chakdah Block towards West, Bangaon Block towards East, GaighataBlock towards East. This Place is on the border of the Nadia District and 24 Parganas North District. 24 Parganas North District Habra-I is South of this place. Bengali is the Local Language here. Total Population of the village is 18019 approximately, where the male population is about 9817 and the female population is about 8502. Public bus service is available within 5 km distance and railway station within 17 km distance.

Table 4.6.4 Cohort study of Kastodanga village

COHORT				
SL No.	Particulars/ Indicators	Years		
		1990-2000	2000-2010	2010-2020
1	Crop Combination	Rice, jute, mustard, brinjal, parwal, chickpea, lentil	Rice, jute, cauliflower, chilli, bitter gourd, flat beans	Flat beans, boro rice(cultivation increased), and jute cultivation decreased and has been replaced by vegetables.
2	Rice Yield (per bigha)	25 mon 1 mon= 40 kg	21-22 mon	18-20 mon
3	Cost of Cultivation of Rice (Rs./bigha)	4000	5500	10000
4	Soil Health	Very good	medium	poor
5	Water Availability	BhomraBil	BhomraBil	Shallow tube well
6	Tree and crop biodiversity	Tree: Shishu, Aswatta, Banyan Rice variety: Jaya, IR50,Ratna, Lal Swarna	Tree: Eucalyptus, Mahogany Rice variety: Ranjit, Pankaj	Tree: Lombu, Kadam Rice variety: Satabdi, Gosai
7	Application of Organic Manure (cow dung)	More	Medium	Very less

8	Livestock Count (cows)	1500	600	30
9	No. of ponds	112	122	135
10	Rainfall (Rainy Days)	High (4-5 months)	Medium (3-4 months)	Less (2 months)
11	Income/Net Return (Rs. /month)	3000	5000	8000
12	Expenditure (Rs./ month)	1000	4000	6500
13	Well-being (Score) (out of 10)	5	7	8

Source: Onefivenine.com and Bharat Chandra Bachar, a farmer of Kastodanga

5

Research Methodology

Research methodology is a detailed plan of investigation and the blueprint of procedure for carrying out the research. In this chapter, a discussion on the methodology has been made to understand the concepts, methods, and techniques, which are utilized to design the study, collect information, analyze data and interpret the findings for the revelation of truth and formulation of theories. The entire chapter has been broken up under the following sub-heads for easy understanding:

1. Locale of research
2. Sampling design
3. Pilot study
4. Variables and their measurements
5. Methods of data collection
6. Statistical tools used for the analysis of data

5.1. Locale of Research

The present study has been conducted in one agro-ecological zones of the state of West Bengal. Villages from Haringhata block in Nadia district were selected from the new alluvial agro-ecological zone to conduct the study. The study was conducted between April 2022 to July 2022.

The characters and the factors under study have been well discernible to this area

- The researcher's close familiarity with respect to area, people, officials, and local dialects
- The ample opportunity to generate relevant data due to the close proximity of the area with the research and extension wing of the State Agricultural Universities;
- The highly cooperative and responsive respondents
- The profuse scope to get relevant information regarding perception on the management of water resources, application of nutrients, improving soil

organic carbon and problems and prospects of practicing conservation agricultural practices

- Experienced, well-versed, venturesome, enthusiast and risk-bearing farmers
- Scope of establishing sustainable and profitable farming opportunities in the area

5.2. Sampling Design

Level	Location	Approach	Justification
State (1)	West Bengal	Purposive	Because of the limitation of time and resources.
Zone (2)	New Alluvial Zone	Purposive	Highly fertile lands and intensive agricultural lands pose threats to soil health declination. That is why the area needs more sustainable approaches like conservation agriculture.
District (1)	Nadia	Purposive	High cropping intensity and chances of high natural resource degradation in the region.
Block (1)	Haringhata	Purposive	These blocks are the most agriculturally advanced blocks experiencing high cropping intensity and degraded soil health.
Villages (4)	Kastodanga Dalilpur Mitrapur Fatehpur	Purposive	Only these were the villages where farmers are growing more crops and were open to discussion about their mental health.
Farmers (100)		Random sampling	Farmers were randomly selected.

Pilot Study

A pilot study has been conducted in the concerned area before constructing the data collecting devices. In course of this survey, informal discussion has been carried out with some farmers, local leaders, extension agencies, and state agricultural officials of the localities. An outline of the socio-economic background of the farmers of the concerned villages, their interest in sustainable agricultural practices, experiences of sustainable farming, and management strategies against the adversities helped in preparing and finalizing the interview schedule.

5.3. Variables and their Measurement

There are two main categories of variables in the present study:

a. Independent variables and

b. Dependent variables

Table 5.1: List of Independent Variables

Independent Variables	Empirical Measurements
Age (x_1)	Chronological age in years
Education (x_2)	Through scoring
Size of land holdings (x_3)	In bigha
Irrigated land (x_4)	In Bigha
Fragments of land (x_5)	In numbers
Family size (x_6)	In numbers
Male female ratio (x_7)	In numbers
Average family education (x_8)	Through scoring
Cropping intensity (x_9)	In percentage
Annual income (x_{10})	In Indian rupees
Communication status (x_{11})	Through scoring and normalization
Food intake of family(x_{12})	In grams
Calorie intake of family (x_{13})	In kcal
Health Status (x_{14})	Through scoring and normalization
Cost of Media Gadgets (x_{15})	In Indian rupees
Cost of Energy Consumption (x_{16})	In Indian rupees
Monthly Cost of smart phone (x_{17})	In Indian rupees
Stress (x_{18})	Through scoring and normalization

The score values have undergone a data normalization process by Z transformation.

a. Independent variables

Age (x_1)

Age is one of the most important determinants of the social status and social role of the individual. Here age refers to the chronological years completed by the respondents at the time of the interview. It has been recorded in integers.

Education (x_2)

Education can be operationalized as the level of formal schooling attained/ literacy acquired by the respondents at the time of the interview. Education plays an important role in changing the behavior and outlook of an individual. It provides knowledge and skill to negotiate the dynamics of social life and grants a unique status in society. Here, the year of education has been considered for study.

Size of land holdings (x_3)

Farm size is a measure of the farm business. Operationally farm size may be defined as a tract of land possessed by an individual to grow crops. In the present study, farming land owned by a family is taken as the measure of farm size.

Irrigated land (x_4)

Irrigated land is measured by measuring how much land is under irrigation.

Fragments of land (x_5)

The number of land fragments is a prime character in South East Asian countries. Cultivation of fragmented lands often ends up being energy and cost prodigal. In this study, it has been operationalized as the disintegration of the total holding into a number of pieces. That is, whether farmers' land is consolidated or disintegrated, if yes, then a total number of disintegrated pieces is taken into account for this study.

Family size (x_6)

Family size has been operationalized as the number of members in an individual's family. In this study, only those members have been considered who were staying together and were sharing meal in one chullah.

Male-female ratio (x_7)

It is the ratio of the total male and females present in the family.

$$\textit{Male female ratio} = \frac{\text{No.of male members}}{\text{No.of female members}}$$

Average family education (x_8)

Education can be operationalized as the level of formal schooling attained/ literacy acquired by the respondents at the time of the interview. Education plays an important role in changing the behavior and outlook of an individual.

It provides knowledge and skill to negotiate the dynamics of social life and grants a unique status in society. Here, the average year of education of all family members has been considered for study.

$$\textit{Average family education} = \frac{\text{Total scores of the family members}}{\text{No.of family members}}$$

Cropping intensity (x_9)

It has been operationalized as the proportion of total annual cropped area by the respondents to their size of operational holdings expressed in percentage. The cropping intensity of an individual farmer has been calculated by using the following formula:

$$\text{Cropping Intensity} = \frac{\text{Gross Cropped Area}}{\text{Net Cropped Area}} x\ 100$$

Annual income (x_{10})

For this study, annual income has been operationalized as the total annual income of the family earned through all the sources and various enterprises in terms of INR.

Communication status (x_{11})

It refers to the use of information source(s) both from outside and within the community for getting agricultural information. The score is assigned as followed:

The information delivered (a)	
Information Received (b)	
Information Shared (c)	

Calculation = 3√a × b × c

Food intake of a family(x_{12})

How many grams of food are taken by the family members?

Caloric intake (x_{13})

Caloric intake is measured by kcal food taken by the farmers in every day.

Health Status (x_{14})

Total health status was measured through a scoring system.

Sl.No	Health problem	The severity of health problem (a)	Frequency of doctor visits (b)	Cost of health care (c)

Calculation = 3√a × b × c

Cost of Media Gadgets (x_{15})

Total Cost of Media Gadgets of the farmer was measured in terms of rupees.

Cost of Energy Consumption (x_{16})

The total family energy consumption of the farmer in a year was measured in terms of rupees.

Monthly Cost of the smartphone (x_{17})

Total Monthly Cost of smartphone of the farmer in a month was measured in terms of rupees.

Stress (x_{18})

Economical stress (a)

- I am stressed because cost per unit is increasing. [ ]
- I am stressed because cost of crop inputs are increasing [ ]
- I am stressed because of my income is uncertain [ ]
- I am stressed because local peoples are migrating for work. [ ]

Psychological stress:- (b)

- I feel stressed by thinking about the future of my farm. [ ]
- I feel stressed by thinking about the loss of traditions. [ ]
- I am stressed because of my family problems. [ ]

Health stress:- (c)

- I am stress because of health problems [ ]
- I am stress because of few health facilities. [ ]

Calculation = $3\sqrt{a \times b \times c}$

The items were finally selected from the expert's rating on 10-point scale.

Table 5.2. List of Dependent Variables

Dependent variables	Notation
Financial Happiness	y_1
Productivity related Happiness	y_2
Family-related Happiness	y_3
Neighbors related Happiness	y_4
Mental health related Happiness	y_5
Weather related Happiness	y_6
Ecology related Happiness	y_7

7 items have been finally selected through judges rating by expert for its final execution in generating responses.

Table 5.3 Financial Happiness (y_1)

Items of financial happiness	Score out of 10
I am happy because my income is more than sufficient. (a)	
I am happy because my income is consistent. (b)	
I feel happy that what I earn is the product of honest and fair work.(c)	
My income is in conformity with my ability and worthiness.(d)	

$$y_1 = \frac{a+b+c+d}{4}$$

Table 5.4. Productivity related Happiness(y_2)

Items of productivity happiness	Score out of 10
I am happy because my farm productivity is more than expected. (a)	
I am happy because I produce the best quality. (b)	
I am happy because my produce has earned a brand. (c)	
I feel happy because my productivity is consistent. (d)	

$$y_2 = \frac{a+b+c+d}{4}$$

Table 5.5. Family related Happiness (y_3)

Items of family happiness	Score out of 10
I am happy because my family is cooperative. (a)	
I feel proud of my family. (b)	
My family always stands by my side during crisis and shares it.(c)	
My decision is always respected by family.(d)	

$$y_3 = \frac{a+b+c+d}{4}$$

Table 5.6. Neighbors related Happiness (y_4)

Items ofneighbors happiness	Score out of 10
I am happy because my neighbors are cooperative. (a)	
Neighbors come helpful through collective wisdom.(b)	
I am happy because my neighbors praise me and my family.(c)	
I am happy because I enjoy free access to my neighbors.(d)	

$$y_4 = \frac{a+b+c+d}{4}$$

Table 5.7. Mental health related Happiness (y_5)

Items mental health happiness	Score out of 10
I am spiritually happy. (a)	
I am happy because I have mental peace. (b)	
I am happy because I have pristine ideas and feelings.(c)	

$$y_5 = \frac{a+b+c}{3}$$

Table 5.8. Weather related Happiness (y_6)

Items ofweather related happiness	Score out of 10
I am happy because weather has been very supportive (a)	
I am happy because bounties of ecology.(b)	
I am happy because climate goes supportive for the last decades.(c)	

$$y_6 = \frac{a+b+c}{3}$$

Table 5.9. Ecology related Happiness (y_7)

Items of ecology related happiness	Score out of 10
I am happy because environment and ecosystem is healthier and productive even today. (a)	
I am happy because my fields are fertile (b)	
I am happy because I can grow any crop in my field at any time. (c)	
I am happy because my lands are so responsive. (d)	

$$y_7 = \frac{a+b+c+d}{4}$$

Preparation of interview schedule

On the basis of finding the pilot study a preliminary schedule was formed with the help of the literature and by the assistance of the Chairman of the Advisory Committee. The interview schedule consisted of three major parts according to the specific objective of the study.

Pre-testing of the interview schedule

It is the process of advance testing of the study design after the schedule has been prepared. The object of pretesting is to detect the discrepancies that have emerged and to remove them after necessary modifications in the schedule. After conducting pretesting appropriate changes and modifications to the interview schedule have been made. The individuals who responded in pretesting have been excluded from the final sample selected for the study.

Techniques of Field Data Collection

Data collection was done by the researcher herself from the farmers of Dalilpur and Kastodanga villages. A telephonic interview based on the interview schedule was conducted. It was depended on Covid-19 situation, socio-political situation and level of responsiveness from the farmers. The contact numbers of the respondent of Kastodanga village were collected from the Kastodanga Gram SamabayKrishiUnnayanSamiti Limited and contact numbers of the respondent of Dalilpur village were collected from few farmers of the village Dalilpur.

Statistical Tools used for Analysis of Data

The statistical methods for analysis and interpretation of raw data are-

a) Range
b) Mean
c) Standard deviation
d) Coefficient of Variance
e) Correlation of coefficient
f) Multiple regression analysis
g) Stepwise regression analysis
h) Path analysis
i) Factor analysis
j) Artificial neural network

Approaches	Purpose	Tools
Descriptive statistics	visualization	Mean, SD,CV
Coefficient of correlation	Relation between two sets of variables	R value
Stepwise regression	Screening of dominant functional variables	R square, Standard error
Path Analysis	Delineating pathway of influence	dY/dX
Canonical Correlation Analysis (CCA)	Conjoint Interpretation of mutual regression between two sets of variables	Cross loading
Artificial Neural Network Analysis(ANN)	Isolation of dominant variables	Inputvariables, output variables, hidden layers
Grouping or constellation of respondents	Homogeneity and distance	Clusteranalysis
Grouping of variables	Grouping pattern of variables based on factor loading	PCA

a) **Range:** The RANGE of a distribution is the difference between the smallest and largest of observations and gives us some idea of the amount of variability present (Panse and Sukhatme, 1967). The range is the most general measure of spread or scatter and may be computed to make a rough comparison of two or more groups for variability. In statistics, the range of a set of data is the difference between the largest and smallest values. It is expressed by the maximum and minimum values of a set of data.

b) **Mean:** The mean is the arithmetic average and is the result obtained when the sum of the values of individuals in the data is divided by the number of individuals in the data (Panse and Sukhatme, 1967). Mean is the simplest and relatively stable measure of central tendency. The mean reflects and is affected by every score in the distribution. Thus, extreme scores affect the mean.

The Mean is computed by the formula

$$\overline{x} = \frac{\sum x}{N}$$

Where, $\overline{x}$ = mean of the scores

$\sum x$ = sum of individual scores

N = number of observations

When the data are expressed in a frequency distribution (grouped), the mean is calculated using the following formula

$$\overline{x} = \frac{\Sigma f \cdot x}{N}$$

Where,

$\overline{x}$ = Mean of the observation.

f = Frequency of the class.

x = Mid value of the class.

N = Number of observations

c) Standard deviation: Standard deviation (SD) of a set of observations is the square root of the arithmetic mean of the squares of all deviations, the deviations being measured from the arithmetic mean of the distribution. It is commonly denoted by the symbol σ (Sigma). It is less affected by sampling error and is a more stable measure of dispersion. The standard deviation of the data grouped in the form of frequency distribution is computed by using the following formula-

$$\sigma = \sqrt{\frac{\Sigma f \cdot d^2}{N}}$$

Where,

f = frequency of the class

d = deviation of the mid-value of the class from the population mean

N = total number of observations

d) Coefficient of Variance: A measure of variation which is independent of the unit of measurement is proved by the coefficient of variation. Being unit free, is useful for the comparison of variability between different populations. The coefficient of variation is the standard deviation expressed as a percentage of the mean. The coefficient of variation is measured by using the following formula —

$$CV= \frac{SD}{Mean}\times 100$$

e) Coefficient of correlation: When increase or decrease in one variety is accompanied by an increase or in the other variants, the two are said to be correlated and the phenomenon is known as CORRELATION. The correlation coefficient (r) is a measure of the relationship between two variables which are at the interval or ratio level of measurement and are linearly related. Pearson product-moment **"r"** is computed by the formula.

$$\mathbf{r}_{xy} = r_{xy} = \frac{N\sum XY - (\sum X)(\sum Y)}{\sqrt{\left[N\sum X^2 - (\sum X)^2\right]\left[N\sum Y^2 - (\sum Y)^2\right]}}$$

When,

X and Y = original scores in variables X and Y

N = number of paired scores

$\sum XY$ = each *X* multiplied by its corresponding *Y*, then summed

$\sum X$ = sum of *X* scores

$\sum X^2$ = each *X* squared, then summed

$(\sum X)^2$ = sum of *X* scores, squared

$\sum Y$ = sum of *Y* scores

$\sum Y^2$ = each *Y* squared, then summed

$(\sum Y)^2$ = sum of *Y* scores, squared

The range of correlation coefficients is between -1 to +1. This means that -1 is a perfect negative correlation, +1 is a perfect positive correlation. If the numbers of errors increase with an increase in typing speed, it indicates a positive correlation and if the numbers of correct words decrease with an increase in typing speed, it is indicative of negative correlation. A correlation coefficient to be acceptable should be statistically significant. Otherwise, we say that no significant relationship exists between the variables.

f) Multiple regression analysis: A multiple correlation coefficient measures the combined relation between a dependent (consequent) and a series of independent (causal) variables. It can also be explained as the correlation between the observed values of the dependent variable and its estimated values from the independent variate values, estimated with the help of the multiple regression equation (Chandel,1970)

If **y** is the consequent variable and **$x_{1,}$** x_2 **....**are the causal variables then the multiple regression equation is given by the following formulas –

$y = a + b_1x_1 + b_2x_2$....

Where, a is a constant and b_1, b_2, ...are the partial regression coefficients.

The square root of the ratio of the regression sum of squares to the total sum of squares is known as the multiple correlation coefficient and is denoted by ***R***. It is always positive and less than one. The square of the multiple correlation coefficient ***R^2*** represents the fraction of the variation in **y**, accounted for by its joint association with the variates **x1** $_{,}x_2$ **....** The coefficient is, therefore a measure of the joint association of all these variates with the dependent variate ***y***, and tells us how much of the variation in ***y*** could be accounted for by reference to these variates (Panse and Sukhatme, 1967).

The beta coefficient is **the degree of change in the outcome variable for every 1-unit of change in the predictor or causal variable**. ... If the beta coefficient is positive, the interpretation is that for every 1-unit increase in the predictor variable, the outcome variable will increase by the beta coefficient value.

g) Stepwise regression analysis: Stepwise regression is a variation of multiple regression which provides a means of choosing independent variables that yield the best prediction possible with the fewest independent variables. It permits the user to solve a sequence of one or more multiple linear regression problems by stepwise application of the least square method. At each step in the analysis, a variable is added or removed which results in the greatest production in the error sum of squares (**Burroughs Corporation**, 1975)

According to **Draper and Smith** (1981), the method of stepwise multiple the analysis is to insert variables in turn until the regression equation is satisfactory. The order of insertion is determined by using the partial correlation coefficient as a measure of the importance of variables not yet in the equation. The program, according to **Burroughs Corporation** (1975), first forms a correlation matrix, finds the best predictor (the independent variable having the highest correlation with the criterion variable) and performs a regression analysis with this predictor. Then. the second best predictor (independent variable having the second highest correlation with the criterion) is found and regression analysis using the multiple correlations of the two best predictors is performed, and so on. At any given step, the group of predictors being used is not necessarily the best group of that size (i.e. the particular group of independent variables does not necessarily have the highest multiple correlations with the criterion that any group of this size does).

Rather, this group contains the variables that have the highest individual correlation with the criterion. Significance of a variable that is being considered for entrance into the regression equation is measured by the F-statistic. If F is too small (less than F 'include'), the variable is not added to the regression equation. Include statement establishes the minimum value of the F-statistic required for the inclusion of a variable in the regression equation. In the example which follows, the F-value for inclusion was 0.01.

Significance of variables already in the regression equation may change as new variables are entered. This significance of the variables currently in the equation is also measured by the F-statistic. If F is too small (less than F 'delete'), the variable is not added to the equation. Delete establishes the value of the F statistic below which the variable is deleted from the regression equation. Here the F-value for deletion was 0.005. The 'tolerance' level specified is used as control degeneracy. Degeneracy, when a variable is entered into the equation, is a linear combination of variables present in the equation. The tolerance statement establishes the maximum value a pivoted element may attain while still allowing its associated variable to be brought into the equation. A variable is not brought into the regression equation if its associated pivoted element is below the specified tolerance level, which was 0.001 in the present example.

h) Path Analysis: The objective of doing Path Analysis is to get a clear picture of the direct and indirect effects of the independent variables on the dependent variable. Variables, through which substantial indirect effects are channeled, are also found out.

Singh and Chaudhary (1977), defined path coefficient as the ratio of the standard deviation of the effect due to a given cause to the total standard deviation of the effect i.e. if **Y** is the effect and **x1**is the cause, the path coefficient for the path from cause **x1** to the effect **Y** is

$\sigma x1/\sigma y$

It is advisable to do path analysis with only those variables which have significant effects on the dependent variable. This may be done by restoring to multiple regression analysis and selecting those independent variables whose partial b values are significant. This shall enhance the clarity of the path analysis.

i) Factor Analysis: Factor analysis is a very useful and popular method of multivariate research technique, mostly used in social and behavioral sciences.

It is a technique applicable when there is a systematic interdependence in finding out something more fundamental or latent which creates this commonality. For instance, we might have data, say, about an individual's income, education, occupation, and dwelling area and want to infer from these some factor (such as social class), which summarizes the commonality of all the said four variables.

Factor analysis seeks to resolve a large set of measured variables in terms of relatively few categories, known as factors. This technique allows the researcher to group variables into factors (based on the correlation between variables), and the factors so derived may be treated as new variables (often termed as latent variables) and their value derived by summing the values of the original variables, which have been grouped into the factor. The meaning and name of such new variable is subjectively determined by the researcher (Kothari,1996).

Since the factors happen to be linear combinations of data, the coordinates of each observation or variable is measured to obtain what are factor loadings. Such factor loading represents the correlation between the particular variable and the factor, and are usually placed in a matrix of correlations between the variables and the factors. In the factor analysis the "Principle Component Analysis" was followed.

j) Artificial Neural Network: Artificial neural networks (**ANNs**), usually simply called **neural networks** (**NNs**), are computing systems inspired by the biological neural networks that constitute animal brains.

An artificial neural network (or simply neural network) consists of an input layer of neurons (or nodes, units), one or two (or even three) hidden layers of neurons, and a final layer of output neurons. Each connection is associated with a numeric number called weight. The output, $\mathbf{h_i}$, of neuron **i**in the hidden layer is,

$$h_i = \sigma \left(\sum_{j=1}^{N} V_{ij} X_j + T_i^{hid}\right)$$

where, σ () = activation (or transfer) function,

N = number of input neurons,

V_{ij} =weights,

X_j = inputs to the input neurons,

T_i^{hid} = threshold terms of the hidden neurons.

The purpose of the activation function is, besides introducing nonlinearity into the neural network, to bound the value of the neuron so that the neural network is not paralyzed by divergent neurons. It has been shown that a neural network constructed the way above can approximate any computable function to arbitrary precision. Numbers given to the input neurons are independent variables and those returned from the output neurons are dependent variables to the function being approximated by the neural network. Inputs to and outputs from a neural network can be binary (such as yes or no) or even symbols (green, red, ...) when data are appropriately encoded. This feature confers a wide range of applicability to neural networks.

The other essential ingredient of a neural network application, namely, *training*. Similar to human learning by examples, a neural network is trained by presenting to it a set of input data called *training set*. The desired outputs of the training data are known so that the aim of the training is to minimize, by adjusting the weights between the connected neurons, *an error function* which is normally the sum of the squared differences between the neural network outputs and the desired outputs.

If we are experimenting with architectures of neural networks, an independent data set called a *validation set* can be applied to the trained neural networks. The one which performs best is then picked up as the one of choice. After validation, yet another independent data set called the *test set* is used to determine the performance level of the neural network which tells how confident we are when using the neural network. It should be understood that a neural network can never learn what is not present in the training set. The size of the training set has therefore to be large enough for the neural network to memorize the features/trends embedded in the training set. On the other hand, if too many unimportant details are contained in the training set, the neural network might waste its resources (weights) fitting the noise. A judicious selection and/or representation of the data are therefore critical to the successful implementations of neural networks.

6

The Case Study

The chapter deals with the Results and discussion of the study, whose purpose is to provide sufficient information so as to arrive at a valid conclusion. Tables and Figures are commonly used in this chapter to support and supplement textual materials. Discussion helps to interpret the Results of the study from a proper perspective and to relate them with other relevant studies.

Table 6.1. Descriptive Statistics of Variable with respect to Minimum, Maximum, Mean, Standard Deviation of Values, Variance, and Coefficient of Variance:

Variable	Minimum	Maximum	Mean	Std. Deviation	Variance	CV (%)
Age (x_1)	25.000	68.000	50.350	9.800	96.048	19.465
Education (x_2)	1.000	15.000	9.220	2.732	7.466	29.636
Size of land holdings (x_3)	0.850	19.820	3.576	2.283	5.211	63.834
Irrigated land (x_4)	0.000	19.000	4.455	3.757	14.112	84.322
Number of land fragments (x_5)	1.000	6.000	1.820	1.009	1.018	55.431
Family size (x_6)	1.000	8.000	4.870	1.315	1.730	27.011
Male female ratio (x_7)	0.250	3.000	1.225	0.594	0.352	48.448
Average family education (x_8)	4.230	11.000	7.783	2.048	4.195	26.317
Cropping intensity (x_9)	1.000	280.000	196.076	75.727	5734.632	38.622
Annual income (x_{10})	48000.000	215000.000	130100.000	32848.443	1079020202.020	25.249
Communication status (x_{11})	0.000	5.430	2.269	1.012	1.023	44.587
Food intake of family(x_{12})	403.000	3784.000	1998.710	643.755	414420.471	32.209
Calorie intake of family(x_{13})	2585.000	20232.000	9727.200	3402.104	11574309.010	34.975
Health status (x_{14})	5.850	36.340	12.021	7.264	52.771	60.430

Variable	Minimum	Maximum	Mean	Std. Deviation	Variance	CV (%)
Cost of media gadgets(x_{15})	0.000	20000.000	7570.000	3400.639	11564343.434	44.923
Cost of energy consumption(x_{16})	300.000	2500.000	729.600	290.676	84492.768	39.841
Monthly cost of a Smartphone (x_{17})	100.000	500.000	287.970	137.840	18999.827	47.866
Stress (x_{18})	2.000	6.650	4.446	1.381	1.909	31.075
Economic happiness (y_1)	3.000	8.500	6.020	1.146	1.313	19.032
Productivity happiness (y_2)	2.500	7.750	5.290	1.089	1.185	20.580
Family happiness (y_3)	4.500	9.250	6.415	1.286	1.654	20.050
Neighbour's happiness (y_4)	3.250	7.750	5.198	1.297	1.681	24.945
Mental health happiness (y_5)	2.330	7.670	5.253	1.250	1.562	23.792
Weather happiness(y_6)	2.000	6.330	4.197	1.001	1.002	23.849
Ecosystem happiness (y_7)	2.500	6.250	4.475	1.175	1.381	26.258

Results: The above table represents the distribution of 18 independent variables (x_1-x_{18}) and 7 dependent variables (y_1-y_7) in terms of their range, mean, standard deviation and co-efficient of variance.

The distribution pattern of the variable, **age (x_1)**, depicts that the minimum age of the respondents is 25 and the maximum is 68. The mean age is 50.350 so most of the respondent belongs to the middle age category with a standard deviation of 9.800. The coefficient of variance (CV) is 19.465 per cent which indicates that the distribution pattern of the variable is highly consistent.

The distribution pattern of the variable, **education** (x_2), shows that the minimum standard of the respondents is 1 and the maximum is 15. The mean education is 9.220 with S.D of 2.732. The coefficient of variance is 29.636 per cent which indicates the distribution of the variables is fairly consistent.

The distribution pattern of the variable, **size of land holding (x_3)**, depicts that the maximum size of holding is 19.820 bigha and the minimum is 0.850bigha. The mean size of holding is 3.576 bigha with a standard deviation of 2.283 bigha. The coefficient of variance is 63.834 per cent which indicates the distribution of the variable is fairly consistent.

The distribution pattern of variable, **irrigated land (x_4)**, depicts that the maximum size of holding is 19 bigha and the minimum is 0.00bigha. The mean size of holding is 4.455 bigha with a standard deviation of 3.757 bigha. The coefficient of variance is 84.322 per cent which indicates the distribution of variable is fairly consistent.

The distribution pattern of variable, **number of land fragments (x_5)**, depicts that the maximum number of fragments is 6 and the minimum is 1. The mean number of fragments is 1.820 with standard deviation of 1.009. The coefficient of variance is 55.431 per cent which indicates the distribution of variables is fairly consistent.

The distribution pattern of variable, **family size (x_6)**, depicts that the maximum family size of the respondents is 8 and the minimum is 1. The mean size of the family is 4.870 with standard deviation of 1.315. The coefficient of variance is 27.011 per cent which indicates the distribution of variables is fairly consistent.

The distribution pattern of the variable, **male female ratio (x_7)**, depicts that the maximum male female ratio of the respondents is 3 and the minimum is 0.250. The mean of the male-female ratio is 1.225 with standard deviation of 0.594. The coefficient of variance is 48.448 per cent which indicates the distribution of variables is fairly consistent.

The distribution pattern of variable, **average family education (x_8),** shows that the minimum standard of the respondents is 4.230 and the maximum is 11. The mean education is 7.783 with S.D of 2.048. The coefficient of variance is 26.317 per cent which indicates distribution of the variables is fairly consistent.

The distribution pattern of variable, **cropping intensity (x_9)**, depicts that the maximum crop intensity is 280 per cent and the minimum is 1 per cent. The mean crop intensity is 196.076 per cent with standard deviation of 75.727 per cent. The coefficient of variance is 38.622 per cent which indicates the distribution of variables is fairly consistent.

The distribution pattern of variable, **annual income (x_{10})**, depicts that the maximum amount annual income is 215000 rupees and the minimum is 48000 rupees. The mean annual income is 130100.00 rupees with standard deviation of 32848.443 rupees. The coefficient of variance is 25.249 per cent which indicates the distribution of variables is fairly consistent.

The distribution pattern of variable, **communication status (x_{11})** depicts that the maximum communication status is 5.430 and the minimum is 0.00. The mean on communication status is 2.269 with standard deviation of1.012. The coefficient of variance is 44.587 per cent which indicates the distribution of variables is fairly consistent.

The distribution pattern of variable, **food intake of family(x_{12})**, depicts that the maximum food intake of family is 3784.00 gm and the minimum is 403.00gm. The mean food intake of family is 1998.710 gm (410gm/ person)with standard deviation of 643.755gm. The coefficient of variance is 32.209 per cent which indicates the distribution of variables is fairly consistent.

The distribution pattern of variable, **calorie intake of family (x_{13})**, depicts that the maximum calorie intake of family is 20232.00 k.cal and the minimum is 2585.00 k cal. The mean on calorie intake of family is 9727.200 kcal (1997 kcal/ person) with standard deviation of 3402.104 k cal. The coefficient of variance is 34.975 per cent which indicates the distribution of variables is fairly consistent.

The distribution pattern of variable, **health status (x_{14})**, depicts that the maximum health status is 36.340 and the minimum is 5.850. The mean on health status is 12.021 with standard deviation of 7.264. The coefficient of variance is 60.430 per cent which indicates the distribution of variable is relatively less consistent.

The distribution pattern of variable, **cost of media gadgets(x_{15})** depicts that the maximum cost of media gadgets is 20000 rupees and the minimum is 00 rupees. The mean cost of media gadgets is 7570 rupees with standard deviation of 3400.639 rupees. The coefficient of variance is 44.923 per cent which indicates the distribution of variables is fairly consistent.

The distribution pattern of variable, **cost of energy consumption(x_{16})**depicts that the maximum cost of energy consumption is 2500 rupees and the minimum is 300 rupees. The mean cost of energy consumption is 729.600 rupees with standard deviation of 290.676 rupees. The coefficient of variance is 39.841 per cent which indicates the distribution of variable is fairly consistent.

The distribution pattern of variable, **monthly Cost of smart phone(x_{17})** depicts that the maximum Monthly Cost of smart phone is 500 rupees and the minimum is 100 rupees. The mean monthly cost of smart phone is 287.970 rupees with standard deviation of 137.840 rupees. The coefficient of variance is 47.866 per cent which indicates the distribution of variable is fairly consistent.

The distribution pattern of variable, **stress (x_{18})**, depicts that the maximum stress is 6.650 and the minimum is 2. The mean on stress is 4.446 with standard deviation of1.381. The coefficient of variance is 31.075per cent which indicates the distribution of variable is fairly consistent.

The distribution pattern of variable, **economic happiness (y_1)**, depicts that the maximum economic happiness is 8.500 and the minimum is 3. The mean of economic happiness is 6.020 with standard deviation of 1.146. The coefficient of variance is 19.032 per cent which indicates the distribution of variable is highly consistent.

The distribution pattern of variable, **productivity happiness (y_2)**, depicts that the maximum productivity happiness is 7.750 and the minimum is 2.500. The mean of productivity happiness is 5.290 with standard deviation of 1.089. The coefficient of variance is 20.580 per cent which indicates the distribution of variable is highly consistent.

The distribution pattern of variable, **family happiness (y_3)**, depicts that the maximum family happiness is 9.250 and the minimum is 4.500. The mean of family happiness is 6.415 with standard deviation of 1.286. The coefficient of variance is 20.050 per cent which indicates the distribution of variable is highly consistent.

The distribution pattern of variable, **neighbours happiness (y_4)**, depicts that the maximum neighbour's happiness is 3.250 and the minimum is 7.750. The mean of neighbours happiness 5.198 with standard deviation of 1.297. The coefficient of variance is 24.945 per cent which indicates the distribution of variable is highly consistent.

The distribution pattern of variable, **mental health happiness (y_5)**, depicts that the maximum Mental health happiness is 7.670 and the minimum is 2.330. The mean of Mental health happiness is 5.253 with standard deviation of 1.250. The coefficient of variance is 23.792per cent which indicates the distribution of variable is highly consistent.

The distribution pattern of variable, **weather happiness (y_6)**, depicts that the maximum Weather happinessis6.330 and the minimum is 2. The mean of Weather happiness is 4.197 with standard deviation of 1.001. The coefficient of variance is 23.849per cent which indicates the distribution of variable is highly consistent.

The distribution pattern of variable, **ecosystem happiness (y_7),** depicts that the maximum Ecosystem happiness is6.250 and the minimum is 2.500. The mean of Ecosystem happiness is 4.475with standard deviation of1.175. The coefficient of variance is 26.258per cent which indicates the distribution of variable is highly consistent.

Table 6.2. Coefficient of Correlation (r): Economic happiness (y_1) Vs. 18 Independent Variables (x_1-x_{18})

Sl.No.	Independent Variables	'r' Value	Remarks
1	Age (x_1)	-0.037	
2	Education (x_2)	**0.225**	*
3	Size of land holdings (x_3)	**0.530**	**
4	Irrigated land (x_4)	-0.156	
5	Number of land fragments (x_5)	-0.082	
6	Family size (x_6)	**0.215**	*
7	Male female ratio (x_7)	-0.166	
8	Average family education (x_8)	**0.338**	**
9	Cropping intensity (x_9)	**0.532**	**
10	Annual income (x_{10})	**0.208**	*
11	Communication status (x_{11})	**0.278**	**
12	Food intake of family(x_{12})	0.174	
13	Calorie intake of family(x_{13})	**0.383**	**
14	Health status (x_{14})	0.183	
15	Cost of media gadgets(x_{15})	**-0.236**	*
16	Cost of energy consumption(x_{16})	**-0.257**	**
17	Monthly cost of Smartphone (x_{17})	**0.319**	**
18	Stress (x_{18})	**-0.340**	**

**Correlation is significant at the 0.01 level

*Correlation is significant at the 0.05 level

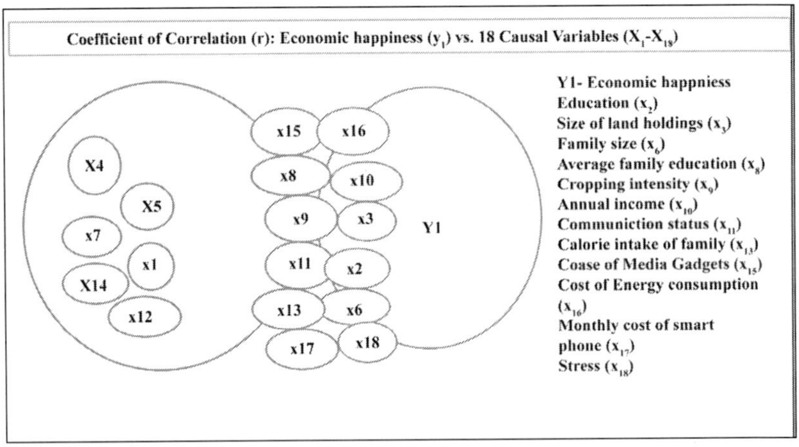

Fig. 6.1. Coefficient of Correlation (r): (y_1) Vs. 18 Independent Variables

Results: Table 6.2 presents the coefficient of correlation between **economic happiness (y_1)** and 18 independent variables. it has been found that the following variables education (x_2)

Size of land holdings (x_3), family size (x_6), average family education (x_8) cropping intensity (x_9)annual income (x_{10})communication status (x_{11}), calorie intake of family (x_{13}) cost of media gadgets (x_{15}) cost of energy consumption (x_{16})monthly cost of smart phone (x_{17})stress (x_{18})have been recorded significant correlation with the dependent variable, economic happiness (y_1).

Revelation: With the up gradation and up scaling of **education** (x_2), the economic happiness has also found to move in the same direction in ecology of entrepreneurial agriculture, education pays dividend.

Land holdings (x_3) offer a resource support to the farm families. It is no wonder that with the higher size of land holding, the economic happiness has been secured.

With the large **family size**(x_6), a family can get a support in the form of family labours to engage in family farm, agri-entrepreneurs, toto (e- rickshaw) drivers and thus choices for income and occupational diversity crop up and more.

Cropping intensity (x_9) indicates diverse and secure crop yield and yield related farm income. So, with higher cropping intensity economic happiness has been escalated.

Annual income(x_{10}) for decisive and obvious reason has positively associated with economic happiness.

When **calorie intake(x_{13})** is on upper echelon for a respondent, the health problem related to poor nutrition go down this is also related with the economic happiness because healthy body is the guarantors for healthy mind and more income.

Cost of media gadgets implies its brand value and quality, at the same time, its cost of maintenance as well these all indicates the intensity of mobile phone usage the happiness moves in the opposite direction this has also been reflected in the correlation direction between **energy** and economic happiness.

It's quite logical that whenever **stress** goes down economic happiness escalates.

Table 6.3. Multiple Regression Analysis: Economic happiness (y_1) vs. 18 Causal Variables (x_1-x_{18})

Sl.No.	Variables	Reg. Coef. B	S.E. B	Beta	t Value
1	Age (x_1)	0.053	0.059	0.053	0.895
2	Education (x_2)	0.234	0.058	0.234	4.022
3	Size of land holdings (x_3)	0.224	0.059	0.224	3.801
4	Irrigated land (x_4)	-0.093	0.087	-0.093	-1.072
5	Number of land fragments (x_5)	0.047	0.084	0.047	0.563
6	Family size (x_6)	0.016	0.058	0.016	0.269
7	Male female ratio (x_7)	0.044	0.052	0.044	0.842
8	Average family education (x_8)	0.148	0.051	0.148	2.875
9	Cropping intensity (x_9)	0.478	0.068	0.478	7.029
10	Annual income (x_{10})	0.134	0.060	0.134	2.212
11	Communication status (x_{11})	0.145	0.054	0.145	2.670
12	Food intake of family(x_{12})	0.119	0.081	0.119	1.467
13	Calorie intake of family(x_{13})	0.261	0.059	0.261	4.395
14	Health status (x_{14})	0.004	0.060	0.004	0.065
15	Cost of media gadgets(x_{15})	-0.144	0.063	-0.144	-2.277
16	Cost of energy consumption(x_{16})	0.010	0.063	0.010	0.153
17	Monthly cost of smartphone (x_{17})	0.058	0.060	0.058	0.972
18	Stress (x_{18})	-0.297	0.073	-0.297	-4.078

R square: 82.10%

The standard error of the estimate: is 0.468

Results: table 6.3 presents the full figure of multiple regression analysis between 18 causal variables (x_1-x_{18}) and economic happiness (y_1).

Revelation: The table 6.3 presents multiple regression analysis (full Figure) between 18 causal variables (x_1-x_{18}) and economic happiness (y_1). The full Figure of multiple regression analysis portraits that with the combination of 18 causal variables, 82.10 per cent variance in economic happiness has been explained.

Cropping Intensity (x_9) has got the highest beta value which is **0.478**

Table 6.4. Stepwise Regression Analysis: Economic happiness (y_1) vs. 18 Causal Variables (x_1-x_{18})

Sl.No.	Variables	Reg.coef. B	S.E. B	Beta	t value
1	Cropping intensity (x_9)	0.497	0.054	0.497	9.165
2	Stress (x_{18})	-0.379	0.053	-0.379	-7.124
3	Size of land holdings (x_3)	0.253	0.053	0.253	4.799
4	Calorie intake of family(x_{13})	0.242	0.049	0.242	4.923
5	Education (x_2)	0.226	0.050	0.226	4.564
6	Media gadgets cost(x_{15})	-0.188	0.049	-0.188	-3.865
7	Communication status (x_{11})	0.141	0.051	0.141	2.793
8	Average family education (x_8)	0.124	0.050	0.124	2.475

R square: 79.60%

The standard error of the estimate: 0.471

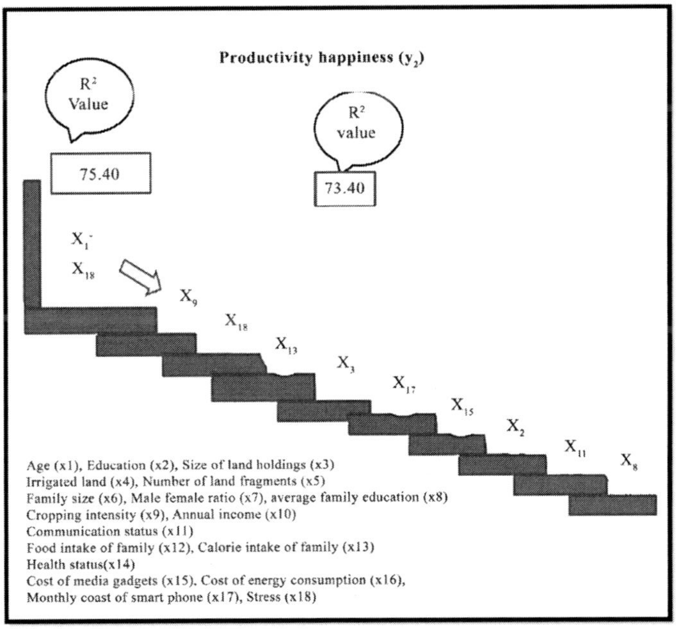

Fig. 6.2: Stepwise Regression Analysis: Economic happiness (y_1) vs. 18 Causal Variables

Results: We have to move for the next table (table 6.4) to isolate the causal dominant variables contributing to the level of economic happiness.

Revelation: It has been found that cropping intensity (x_9),stress (x_{18}),size of land holdings (x_3)calorie intake of family(x_{13}),education (x_2) ,media gadgets cost(x_{15}),communication status(x_{11}),average family education (x_8) have been retained at the last step and has substantially contributed to the consequent variable economic happiness (y_1).

- The higher is the **cropping intensity,** better has been the happiness.
- Similarly, lesser the **stress,** higher has been economic happiness.
- The higher is the **landholdings,** better has been the happiness.
- The higher is the **calorie intake of family,** better has been the happiness.
- The higher is the **education,** better has been the happiness.
- The lesser is the **media gadgets cost,** better has been the happiness.
- The higher is the **communication status,** better has been the happiness.
- The higher is the **average family education,** better has been the happiness.

These 8 variables together have contributed 79.60% of variance i.e. $\left(\frac{79.60}{82.10}\times 100\right) = 97.95$. The net contribution made by the 8 dominant causal variables stands at 97.95 per cent.

So, the eight variables have got formidable causal impact.

Table 6.5. Path Analysis: Decomposition of Total Effect into Direct, Indirect and Residual Effect: Economic happiness (y1) vs. 18 exogenous variables (x_1-x_{18})

Sl.No	Variables	Total Effect	Direct Effect	Indirect Effect	Highest Indirect Effect
1	Age (x1)	-0.037	0.052	-0.089	-0.112 (x_2)
2	Education (x_2)	0.225	0.232	-0.007	0.046 (x_9)
3	Size of land holdings (x3)	0.530	0.223	0.307	0.1 (x_9)
4	Irrigated land (x4)	-0.156	-0.094	-0.062	-0.053 (x_9)
5	Number of land fragments (x5)	-0.082	0.048	-0.130	-0.076 (x_4)
6	Family size (x6)	0.215	0.017	0.198	0.096 (x_{18})
7	Male female ratio (x7)	-0.166	0.044	-0.210	-0.069 (x_9)
8	Average family education (x8)	0.338	0.148	0.190	0.063 (x_9)
9	Cropping intensity (x9)	0.532	0.478	0.054	-0.084 (x_{18})
10	Annual income (x10)	0.208	0.134	0.074	0.100 (x_9)
11	Communication status (x11)	0.278	0.145	0.133	0.093 (x_9)
12	Food intake of family(x12)	0.174	0.117	0.057	-0.182 (x_9)
13	Calorie intake of family(x13)	0.383	0.262	0.121	0.05 (x_{18})
14	Health status (x14)	0.183	0.003	0.180	0.125 (x_{18})
15	Cost of media gadgets(x15)	-0.236	-0.146	-0.090	0.125 (x_{18})
16	Cost of energy consumption(x16)	-0.257	0.011	-0.268	-0.117 (x_9)
17	Monthly cost of smartphone (x17)	0.319	0.059	0.260	0.124 (x_9)
18	Stress (x18)	-0.340	-0.297	-0.043	0.135 (x_9)

Residual effect: 0.179

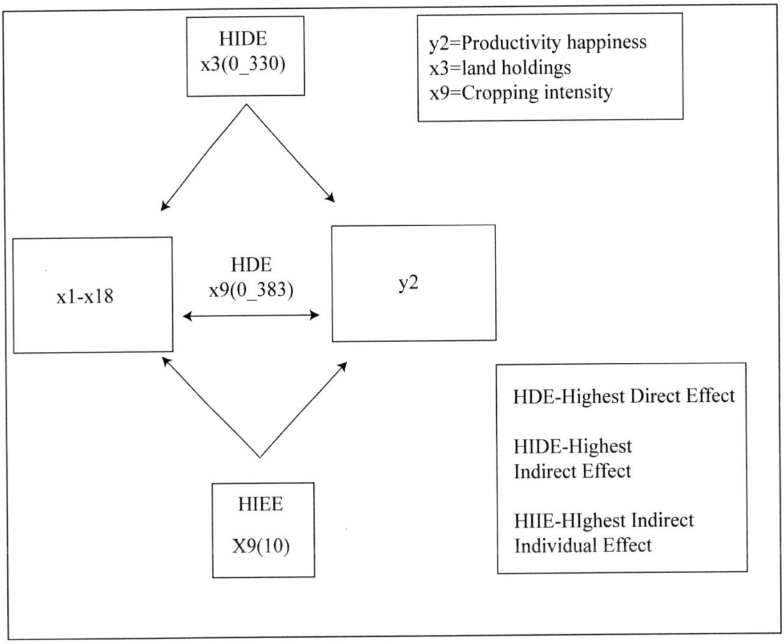

Fig. 6.3. Path Analysis: Decomposition of Total Effect into Direct, Indirect and Residual Effect: Economic happiness ($y_{1)}$ vs. 18 exogenous variables (x_1-x_{18})

Results: Table 6.5 presents the path analysis wherein the total effect of exogenous variable on consequent variable has been decomposed into direct, indirect and residual effect.

Revelation: It has been found that highest direct effect has been exerted by causal variable cropping intensity and the highest indirect effect has been exerted by size of land holdings (x3). And interestingly it has routed the highest indirect effect of 11 other causal variables to ultimately characterized economic happiness.

Table 6.6. Coefficient of Correlation (r): Productivity happiness (y_2) Vs. 18 Independent Variables

S.No.	Independent Variables	'r' Value	Remarks
1	Age (x1)	-0.036	
2	Education (x_2)	0.184	
3	Size of land holdings (x3)	0.464	**
4	Irrigated land (x4)	-0.161	
5	Number of land fragments (x5)	-0.101	
6	Family size (x6)	0.189	

S.No.	Independent Variables	'r' Value	Remarks
7	Male female ratio (x7)	-0.223	*
8	Average family education (x8)	0.302	**
9	Cropping intensity (x9)	0.479	**
10	Annual income (x10)	0.113	
11	Communication status (x11)	0.264	**
12	Food intake of family(x12)	0.220	*
13	Calorie intake of family(x13)	0.420	**
14	Health status (x14)	0.241	*
15	Cost of media gadgets(x15)	-0.227	*
16	Cost of energy consumption(x16)	-0.247	*
17	Monthly cost of smartphone (x17)	0.428	**
18	Stress (x18)	-0.318	**

**Correlation is significant at the 0.01 level

*Correlation is significant at the 0.05 level

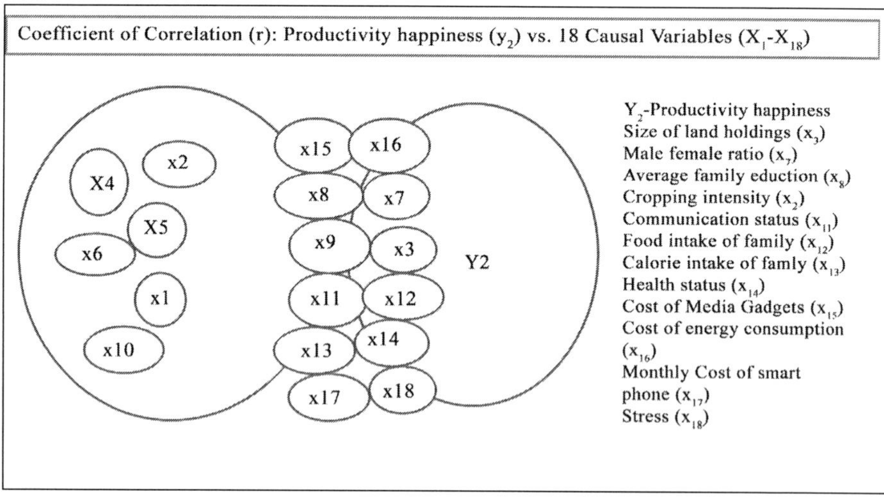

Fig. 6.4. Coefficient of Correlation (r): Productivity happiness (y_2) Vs. 18 Independent Variables

Results: Table-6.6 presents the coefficient of correlation between **Productivity happiness (y_1)** and 18 independent variables. It has been found that the following variables Size of land holdings (x_3)Male female ratio (x_7)Average family education (x_8), Cropping intensity (x_9), Communication status (x_{11}), Food intake of family(x_{12}), Calorie intake of family (x_{13}), Health Status (x_{14}), cost of Media Gadgets (x_{15}), cost of Energy Consumption (x_{16}), Monthly Cost of smart phone (x_{17}), stress (x_{18}) have recorded significant correlation with the dependent variable, productivity happiness (y_2).

Revelation: With the up gradation and up scaling of **land holdings** (x_3), the productivity related happiness has also found to move in the same direction, because when land holding is higher the productivity will also increase. So, the productivity related happiness will also increase.

Male female ratio (x_7) has recorded negative correlation with the productivity happiness.(y_2) It indicates the feeble participation of women in the agriculture production process at large so there is ample scope for mobilization of women into the agricultural production process and its subsequent management.

With the up gradation of **average family education (x_8)** the productivity related to happiness (y_2) will also increase. In ecology of entrepreneurial agriculture, the average family education pays dividends.

Cropping intensity (x_9) indicates diverse and secure crop yield, so with the higher cropping intensity productivity related happiness has been escalated. So, cropping intensity has a positive correlation with the productivity happiness (y_2).

Communication status (x_{11}) implies the wider and faster communication if communication is more the input related information will also be more so, it indicates more productivity so productivity related happiness (y_2) will be more.

Food intake of family(x_{12}), the better is the food intake the better is the corporal ability to unleash labour.

Calorie intake of family (x_{13}), the better is the food intake the higher will be the calorie intake and **health status (x_{14})** will also be fine so the better will be the labour proficiency.

Cost of Media Gadgets (x_{15}) and **energy consumption (x_{16})** has a negative correlation with productivity happiness(y_2). The farmers are using vigorous amount of fertilizer it effects the farm productivity and yield.

Smart phone(x_{17}) has a positive impact on productivity related happiness. Higher is the usage of smart phone, frequent and wider has been the network.

It is quite logical that whenever stress(x18) goes down economic happiness escalates.

Table 6.7. Multiple Regression Analysis: Productivity happiness (y_2) Vs. 18 Causal Variables

Sl.No.	Variables	Reg. Coef. B	S.E. B	Beta	t Value
1	Age (x1)	0.031	0.069	0.031	0.442
2	Education (x_2)	0.207	0.068	0.207	3.051
3	Size of land holdings (x3)	0.134	0.069	0.134	1.945
4	Irrigated land (x4)	-0.040	0.102	-0.040	-0.396
5	Number of land fragments (x5)	0.001	0.098	0.001	0.012
6	Family size (x6)	0.040	0.068	0.040	0.595
7	Male female ratio (x7)	-0.012	0.061	-0.012	-0.190
8	Average family education (x8)	0.146	0.060	0.146	2.425
9	Cropping intensity (x9)	0.385	0.080	0.385	4.829
10	Annual income (x10)	0.057	0.071	0.057	0.808
11	Communication status (x11)	0.166	0.063	0.166	2.622
12	Food intake of family(x12)	0.115	0.095	0.115	1.212
13	Calorie intake of family(x13)	0.308	0.069	0.308	4.441
14	Health status (x14)	0.120	0.070	0.120	1.709
15	Cost of media gadgets(x15)	-0.162	0.074	-0.162	-2.180
16	Cost of energy consumption(x16)	0.014	0.074	0.014	0.189
17	Monthly Cost of smartphone (x17)	0.225	0.070	0.225	3.215
18	Stress (x18)	-0.145	0.085	-0.145	-1.707

R square: 75.40%

The standard error of the estimate: 0.548

Results: Table 6.7 presents the full model of multiple regression analysis between 18 causal variables (x_1-x_{18}) and Productivity happiness (y_2).

Revelation: The table-6.7 presents multiple regression analysis (full model) between 18 causal variables (x_1-x_{18}) and productivity happiness (y_2).The full Figure of multiple regression analysis portraits that with the combination of 18 causal variables, 75.40 percent variance in productivity happiness (y_2).has been explained.

Cropping Intensity (x_9) has got the highest beta value which is 0.385

Table 6.8. Stepwise Regression Analysis: Productivity happiness (y_2) Vs. 18 Causal Variables

Sl.No	Variables	Reg.coef. B	S.E. B	Beta	t value
1	Cropping intensity (x9)	0.344	0.067	0.344	5.115
2	Stress (x18)	-0.246	0.066	-0.246	-3.722
3	Calorie intake of family(x13)	0.322	0.057	0.322	5.665
4	Size of land holdings (x3)	0.191	0.061	0.191	3.150
5	Monthly Cost of smart phone(x17)	0.247	0.062	0.247	3.991
6	Media gadgets cost(x15)	-0.183	0.056	-0.183	-3.276

Sl.No	Variables	Reg.coef. B	S.E. B	Beta	t value
7	Education (x_2)	0.187	0.057	0.187	3.285
8	Communication status (x11)	0.174	0.059	0.174	2.950
9	Average family education (x8)	0.133	0.058	0.133	2.320

R square: 73.40%

The standard error of the estimate: 0.541

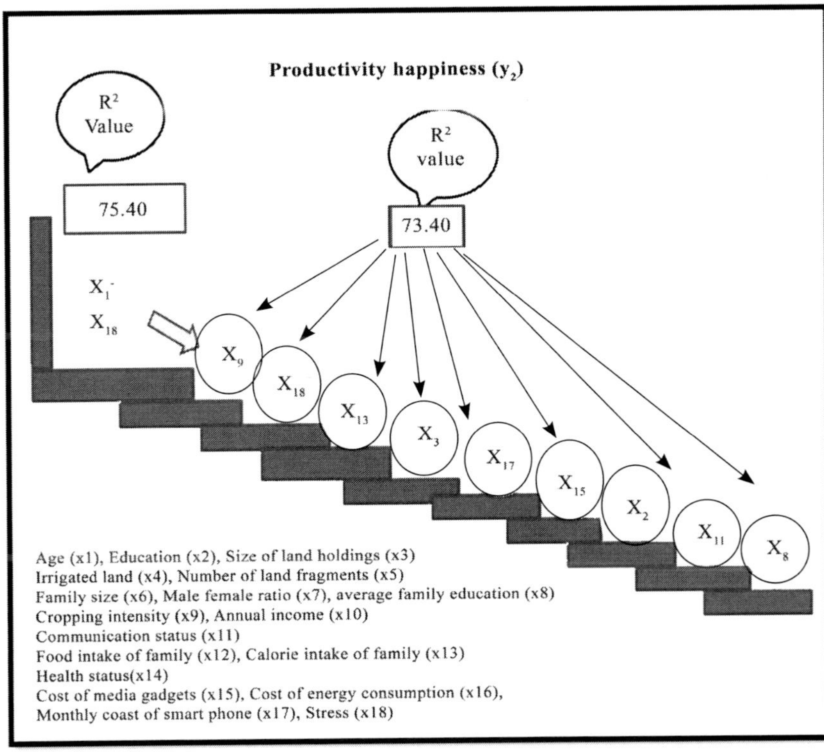

Fig. 6.5. Stepwise Regression Analysis: Productivity happiness (y_2) Vs. 18 Causal Variables (x_1-x_{18})

Results: We have to move for the next table (table 6.8) to isolate the causal dominant variables contributing to the level of productivity happiness.

Revelation: It has been found that cropping intensity (x_9), stress (x_{18}), calorie intake of family(x_{13}), size of land holdings (x_3) monthly cost of smart phone (x17)media gadgets cost(x_{15})education (x_2) communication status (x_{11})average family education (x_8) have been retained at the last step and has substantially contributed to the consequent variable productivity happiness (y_2).

- The higher is the cropping intensity (x_9) the better productivity happiness.
- The higher is the stress (x_{18}) the lower has been the productivity happiness.
- The higher is the calorie intake of family (x_{13}) the better has been productivity happiness.
- The higher is the size of land holdings (x_3) the better has been productivity happiness.
- The higher is the monthly cost of smart phone (x_{17}) the better has been productivity happiness.
- The higher is the media gadgets cost (x_{15}), the lower has been the productivity happiness.
- The higher is the education (x_2), the better has been productivity happiness.
- The higher is the communication status (x_{11}) the better has been productivity happiness.
- The higher is the average family education (x_8), the better has been productivity happiness.

These 9 variables together have contributed 73.40% of variance i.e. (= 97.34 The net contribution made by the 9 dominant causal variables stands at 97.34 per cent. So the 9 variables have got formidable causal coefficient.

Table 6.9. Path Analysis: Decomposition of Total Effect into Direct, Indirect and Residual Effect: productivity happiness y2 vs. 18 exogenous variables

Sl.No	Variables	Total Effect	Direct Effect	Indirect Effect	Highest Indirect Effect
1	Age (x1)	-0.036	0.032	-0.068	-0.1 (x_2)
2	Education (x_2)	0.184	0.208	-0.024	0.037 (x9)
3	Size of land holdings (x3)	0.464	0.134	0.330	0.08 (x9)
4	Irrigated land (x4)	-0.161	-0.042	-0.119	-0.057 (x15)
5	Number of land fragments (x5)	-0.101	0.003	-0.104	-0.056 (x15)
6	Family size (x6)	0.189	0.041	0.148	0.047 (x18)
7	Male female ratio (x7)	-0.223	-0.012	-0.211	-0.056 (x9)
8	Average family education (x8)	0.302	0.146	0.156	0.05 (x9)
9	Cropping intensity (x9)	0.479	0.383	0.096	0.059 (x17)
10	Annual income (x10)	0.113	0.057	0.056	-0.106 (x13)
11	Communication status (x11)	0.264	0.167	0.097	0.075 (x9)
12	Food intake of family(x12)	0.220	0.113	0.107	-0.147 (x9)
13	Calorie intake of family(x13)	0.420	0.309	0.111	0.037 (x12)
14	Health status (x14)	0.241	0.120	0.121	-0.074 (x9)
15	Cost of media gadgets(x15)	-0.227	-0.164	-0.063	-0.024 (x3)
16	Cost of energy consumption(x16)	-0.247	0.016	-0.263	-0.094 (x9)
17	Monthly Cost of smart phone(x17)	0.428	0.227	0.201	0.100 (x9)
18	Stress (x18)	-0.318	-0.145	-0.173	0.108 (x9)

Residual effect: 0.245

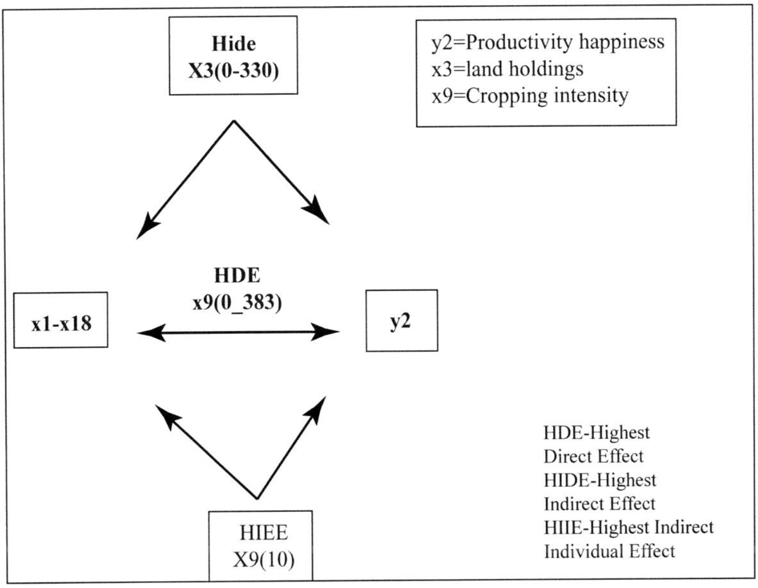

Fig. 6.6. Path Analysis: Decomposition of Total Effect into Direct, Indirect and Residual Effect: productivity happiness y2 vs. 18 exogenous variables (x_1-x_{18})

Results: Table 6.9 presents the path analysis wherein the total effect of exogenous variable on consequent variable has been decomposed into direct, indirect and residual effect.

Revelation: It has been found that highest direct effect has been exerted by causal variable cropping intensity (x_9) and the highest indirect effect has been exerted by Size of land holdings (x_3). And interestingly it has routed the highest indirect effect of 10 other causal variables to ultimately characterized productivity happiness (y_2).

Table 6.10. Coefficient of Correlation (r): family happiness (y_3) Vs. 18 Independent Variables (x_1-x_{18})

Sl.No.	Independent Variables	'r' Value	Remarks
1	Age (x_1)	0.029	
2	Education (x_2)	0.199	*
3	Size of land holdings (x_3)	0.426	**
4	Irrigated land (x_4)	-0.163	
5	Number of land fragments (x_5)	-0.117	
6	Family size (x_6)	0.211	*
7	Male female ratio (x_7)	-0.218	*
8	Average family education (x_8)	0.214	*

9	Cropping intensity (x_9)	0.390	**
10	Annual income (x_{10})	0.128	
11	Communication status (x_{11})	0.232	*
12	Food intake of family (x_{12})	0.370	**
13	Calorie intake of family(x_{13})	0.261	**
14	Health status (x_{14})	0.179	
15	Cost of media gadgets (x_{15})	-0.156	
16	Cost of energy consumption (x_{16})	-0.170	
17	Monthly Cost of smart phone (x_{17})	0.509	**
18	Stress (x_{18})	-0.543	**

**Correlation is significant at the 0.01 level

*Correlation is significant at the 0.05 level

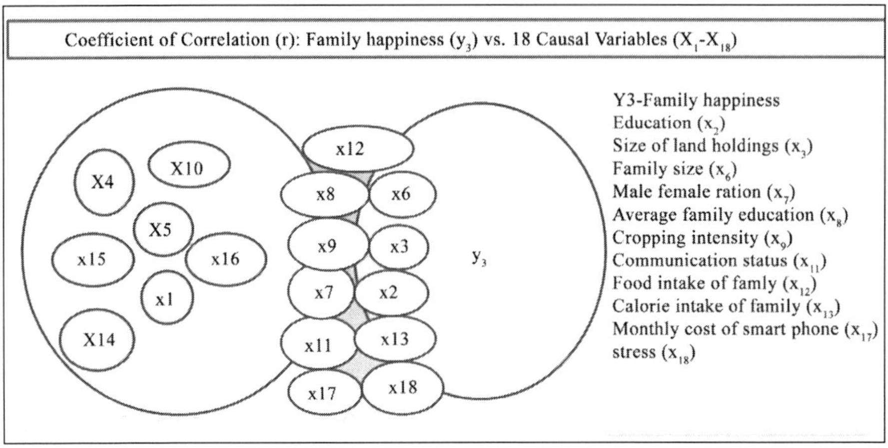

Fig. 6.7. Coefficient of Correlation (r): family happiness (y_3) Vs. 18 Independent Variables

Results: Table-6.10 presents the coefficient of correlation between **family happiness (y_3)** and 18 independent variables. it has been found that the following variables education (x_2)size of land holdings (x_3)family size (x_6) male female ratio (x_7) average family education (x_8)cropping intensity (x_9) communication status (x_{11})food intake of family(x_{12})calorie intake of family (x_{13}). Monthly Cost of smart phone (x_{17}) stress (x_{18}) has recorded significant correlation with the dependent variable, family happiness (y_3).

Revelation: With the up gradation and up scaling of **education**(x_2), the family happiness (y_3) has also found to move in the same direction in ecology of entrepreneurial agriculture education pays dividend. Family happiness will increase with increasing education.

Land holdings (x_3) , offer a resource support to the farm families. It is no wander that with the higher size of land holding, the family happiness has been secured.

With the large **family size**(x_6), a family can get a support in the form of family labours engage in family farm, agri-entrepreneurs , toto (e- rickshaw) drivers and thus family happiness has been secured.

Gender ratio (x_7) , has got a decisive contribution to both family and happiness. The ground reality reveals that family nurturing more female member's perception of happiness takes some issues and stresses. It can be considered as a lower level of happiness due to some stereotypes and wrong visualization.

With the up gradation and up scaling of **average family education(x_8),** the family happiness (y_3) has also found to move in the same direction in ecology of entrepreneurial agriculture education pays dividend. Family happiness will increase with increasing education.

Cropping intensity (x_9), indicates diverse and secure crop yield and yield related farm income. So, with higher cropping intensity family happiness has been escalated.

Communication status (x_{11}), implies the wider and faster communication if communication is more the input related information will also be more so, it indicates more communication with the family members as well so family happiness(y_3) will be more.

Food intake of family(x_{12}), the better is the food intake the better is the corporal ability to unleash labour.

Calorie intake of family (x_{13}), the better is the food intake the higher will be the calorie intake will also be fine so the better will be the labour proficiency and better will be the family happiness.

Smart phone(x_{17}) has a positive impact on family related happiness. Higher is the usage of smart phone, frequent and wider has been the network so family will be closer.

It's quite logical that whenever stress(x_{18}), goes down family happiness escalates.

Table 6.11. Multiple Regression Analysis: Family happiness (y_3) Vs. 18 Causal Variables

Sl.No.	Variables	Reg. Coef. B	S.E. B	Beta	t Value
1	Age (x_1)	0.114	0.063	0.114	1.809
2	Education (x_2)	0.217	0.062	0.217	3.510
3	Size of land holdings (x_3)	0.173	0.063	0.173	2.748
4	Irrigated land (x_4)	-0.080	0.092	-0.080	-0.871
5	Number of land fragments (x_5)	0.046	0.089	0.046	0.512
6	Family size (x_6)	-0.006	0.062	-0.006	-0.090
7	Male female ratio (x_7)	0.008	0.056	0.008	0.152
8	Average family education (x_8)	0.076	0.055	0.076	1.396
9	Cropping intensity (x_9)	0.407	0.072	0.407	5.614
10	Annual income (x_{10})	0.098	0.064	0.098	1.528
11	Communication status (x_{11})	0.110	0.058	0.110	1.914
12	Food intake of family(x_{12})	0.190	0.086	0.190	2.212
13	Calorie intake of family(x_{13})	0.100	0.063	0.100	1.580
14	Health status (x_{14})	-0.068	0.064	-0.068	-1.068
15	Cost of media gadgets(x_{15})	-0.095	0.067	-0.095	-1.404
16	Cost of energy consumption(x_{16})	0.017	0.067	0.017	0.256
17	Monthly Cost of smart phone(x_{17})	0.216	0.064	0.216	3.402
18	Stress (x_{18})	-0.466	0.077	-0.466	-6.021

R square: 79.70 %

The standard error of the estimate: 0.498

Results: Table6.11 presents the full model of multiple regression analysis between 18 causal variables (x_1-x_{18}) and family happiness (y_3).

Revelation: The table-6.11 presents multiple regression analysis (full model) between 18 causal variables (x_1-x_{18}) and family happiness (y_3).The full Figure of multiple regression analysis portraits that with the combination of 18 causal variables, 79.70 percent variance in family happiness (y_3) have been explained.

Cropping intensity (x_9) has got the highest beta value. The value is 0.407

Table 6.12. Stepwise Regression Analysis: Family happiness (y_3) vs. 18 Causal Variables

Sl.No	Variables	Reg.coef. B	S.E. B	Beta	t value
1	Stress (x_{18})	-0.457	0.067	-0.457	-6.855
2	Cropping intensity (x_9)	0.483	0.064	0.483	7.519
3	Food intake of family(x_{12})	0.237	0.068	0.237	3.491
4	Size of land holdings (x_3)	0.192	0.054	0.192	3.538
5	Education (x_2)	0.182	0.051	0.182	3.534
6	Monthly Cost of smart phone(x_{17})	0.172	0.058	0.172	2.950
7	Communication status (x_{11})	0.116	0.054	0.116	2.139
8	Media gadgets cost(x_{15})	-0.108	0.052	-0.108	-2.062

R square: 77.30%

The standard error of the estimate: 0.497

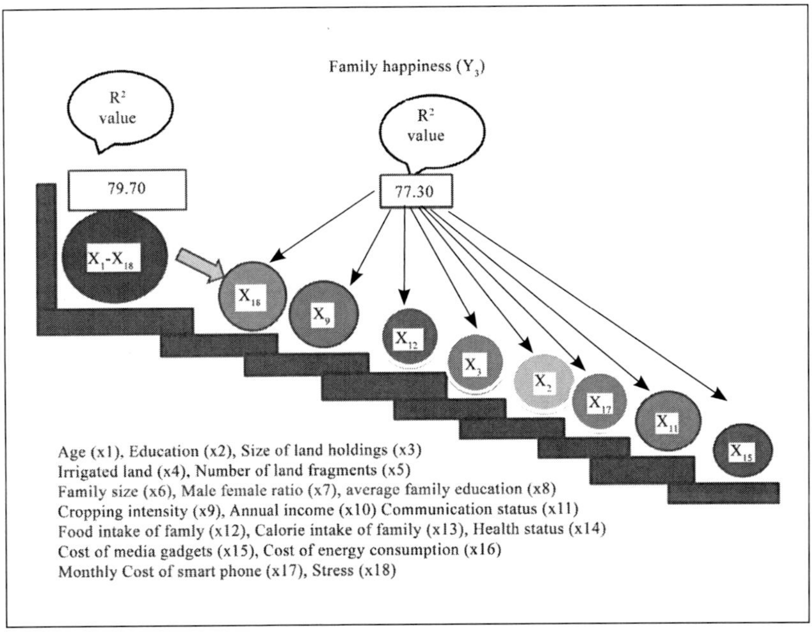

Fig. 6.8. Stepwise Regression Analysis: Family happiness (y_3) vs. 18 Causal Variables

Results: We have to move for the next table (table 6.12) to isolate the causal dominant variables contributing to the level of family happiness.

Revelation: It has been found that stress (x_{18}), cropping intensity (x_9), food intake of family(x_{12}), size of land holdings (x_3), education (x_2), monthly cost of smart phone(x_{17}), communication status (x_{11}), media gadgets cost (x_{15}), have been retained at the last step and has substantially contributed to the consequent variable family happiness (y_3).

- The higher is the stress (x_{18}), lower would be the family happiness (y_3).
- The higher is the cropping intensity (x_9), higher would be the family happiness (y_3).
- The higher is the food intake of family(x_{12}), higher would be the family happiness (y_3).
- The higher is the size of land holdings (x_3), higher would be the family happiness (y_3).
- The higher is the education (x_2), higher would be the family happiness (y_3).

- The higher the monthly cost of smart phone (x_{17}), higher will be the family happiness (y_3).
- The higher the communication status (x_{11}), higher will be the family happiness (y_3).
- The higher the media gadgets cost (x_{15}), lower will be the family happiness (y_3).

These 8 variables together have contributed 77.30% of variance i.e. $\left(\frac{77.30}{79.70}\times 100\right)$ = 96.98 The net contribution made by the 8 dominant causal variables stands at 96.98 percent. So these eight variables have got formidable causal impact.

Table 6.13. Path Analysis: Decomposition of Total Effect into Direct, Indirect and Residual Effect: Family Happiness y3 vs. 18 exogenous variables

Sl.No	Variables	Total Effect	Direct Effect	Indirect Effect	Highest Indirect Effect
1	Age (x_1)	0.029	0.114	-0.085	-0.104 (x_2)
2	Education (x_2)	0.199	0.216	-0.017	-0.055 (x_1)
3	Size of land holdings (x_3)	0.426	0.173	0.253	0.085 **(x_{18})**
4	Irrigated land (x_4)	-0.163	-0.081	-0.082	-0.045 (x_9)
5	Number of land fragments (x_5)	-0.117	0.046	-0.163	-0.065 (x_4)
6	Family size (x_6)	0.211	-0.005	0.216	0.15 **(x_{18})**
7	Male female ratio (x_7)	-0.218	0.009	-0.227	-0.064 **(x_{18})**
8	Average family education (x_8)	0.214	0.076	0.138	0.053 (x_9)
9	Cropping intensity (x_9)	0.390	0.407	-0.017	-0.131 **(x_{18})**
10	Annual income (x_{10})	0.128	0.099	0.029	0.085 (x_9)
11	Communication status (x_{11})	0.232	0.110	0.122	0.079 (x_9)
12	Food intake of family(x_{12})	0.370	0.191	0.179	0.27 **(x_{18})**
13	Calorie intake of family(x_{13})	0.261	0.100	0.161	0.078 **(x_{18})**
14	Health status (x_{14})	0.179	-0.068	0.247	0.196 **(x_{18})**
15	Cost of media gadgets(x_{15})	-0.156	-0.095	-0.061	-0.031 (x_3)
16	Cost of energy consumption(x_{16})	-0.170	0.018	-0.188	-0.1 00(x_9)
17	Monthly Cost of smart phone(x_{17})	0.509	0.217	**0.292**	0.118 **(x_{18})**
18	Stress (x_{18})	**-0.543**	**-0.465**	-0.078	0.115 (x_9)

Residual effect: 0.203

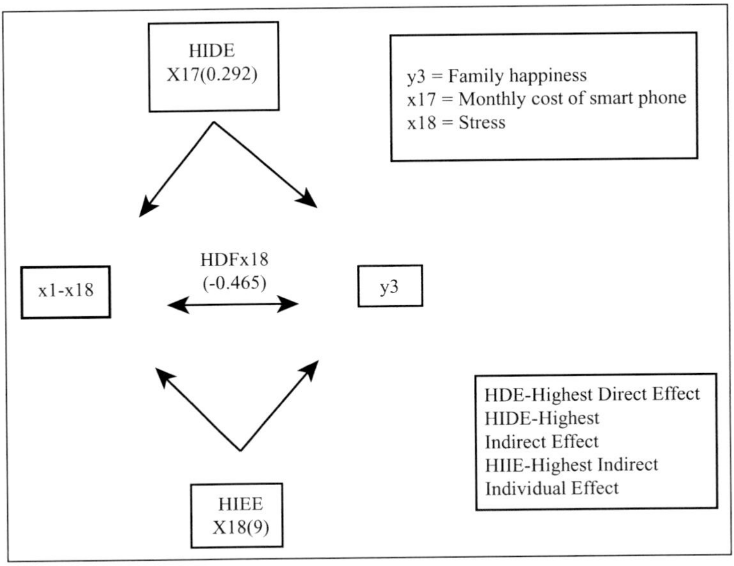

Fig. 6.9. Path Analysis: Decomposition of Total Effect into Direct, Indirect and Residual Effect: y_3 vs. 18 exogenous variables

Results: Table 6.13 presents the path analysis wherein the total effect of exogenous variable on consequent variable has been decomposed into direct, indirect and residual effect.

Revelation: It has been found that highest direct effect has been exerted by causal variable stress (x_{18}), and the highest indirect effect has been exerted by cost of smart phone (x_{17}). And interestingly it has routed the highest indirect effect of 8 other causal variables to ultimately characterized family happiness (y_3).

Table 6.14. Coefficient of Correlation (r): Neighbours Happiness (y_4) Vs. 18 Independent Variables

Sl.No.	Independent Variables	'r' Value	Remarks
1	Age (x_1)	0.047	
2	Education (x_2)	0.117	
3	Size of land holdings (x_3)	0.347	**
4	Irrigated land (x_4)	-0.061	
5	Number of land fragments (x_5)	0.014	
6	Family size (x_6)	0.178	
7	Male female ratio (x_7)	-0.100	
8	Average family education (x_8)	0.108	
9	Cropping intensity (x_9)	-0.019	

Sl.No.	Independent Variables	'r' Value	Remarks
10	Annual income (x_{10})	-0.030	
11	Communication status (x_{11})	0.198	*
12	Food intake of family(x_{12})	0.491	**
13	Calorie intake of family(x_{13})	0.356	**
14	Health status (x_{14})	0.456	**
15	Cost of media gadgets(x_{15})	-0.017	
16	Cost of energy consumption(x_{16})	-0.084	
17	Monthly Cost of smart phone(x_{17})	0.339	**
18	Stress (x_{18})	-0.654	**

**Correlation is significant at the 0.01 level

*Correlation is significant at the 0.05 level

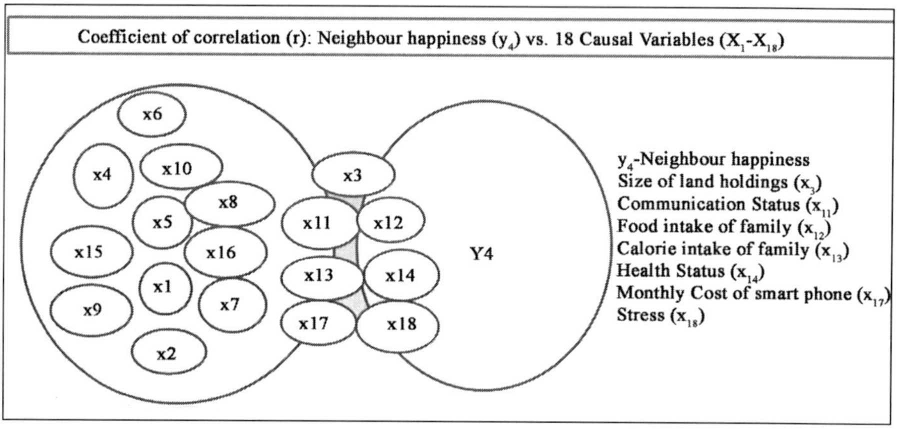

Fig. 6.10. Coefficient of Correlation (r): Neighbours Happiness (y_4) Vs. 18 Independent Variables

Results: Table-6.14 presents the coefficient of correlation between **Neighbours happiness (y_4)** and 18 independent variables. It has been found that the following variables Size of land holdings (x_3), Communication status (x_{11}), Food intake of family(x_{12}), Calorie intake of family (x_{13}), Health Status (x_{14}), Monthly Cost of smart phone (x_{17}), stress (x_{18}) have recorded significant correlation with the dependent variable, neighbours happiness (y_4).

Revelation: Land holdings (x_3), offer a resource support to the farm families as well as their neighbours. It is no wander that with the higher size of land holding, the neighbour's happiness has been secured.

Communication status (x_{11}), implies the wider and faster communication if communication is more the input related information will also be more also it indicates more communication with the family members and neighbors. So neighbor's happiness (y_4) will escalate.

Food intake of family(x_{12}), the better is the food intake the better is the corporal ability to unleash labour.

Calorie intake of family (x_{13}), the better is the food intake the higher will be the calorie intake and **health status** (x_{14}), will also be fine so the better will be the labour proficiency and better will be the neighbor happiness.

Smart phone(x_{17}), has a positive impact on neighbour related happiness. Higher is the usage of smart phone, frequent and wider has been the network so neighbour will be closer.

It's quite logical that whenever **stress(x_{18})**, goes down neighbour happiness escalates.

Table 6.15 Multiple Regression Analysis: Neighbour happiness (y_4) vs. 18 Causal Variables

Sl.No.	Variables	Reg. Coef. B	S.E. B	Beta	t Value
1	Age (x_1)	0.100	0.081	0.100	1.236
2	Education (x_2)	0.225	0.079	0.225	2.841
3	Size of land holdings (x_3)	0.137	0.080	0.137	1.706
4	Irrigated land (x_4)	-0.112	0.118	-0.112	-0.943
5	Number of land fragments (x_5)	0.137	0.114	0.137	1.194
6	Family size (x_6)	-0.067	0.079	-0.067	-0.850
7	Male female ratio (x_7)	0.104	0.072	0.104	1.460
8	Average family education (x_8)	0.020	0.070	0.020	0.293
9	Cropping intensity (x_9)	0.049	0.093	0.049	0.532
10	Annual income (x_{10})	0.017	0.082	0.017	0.201
11	Communication status (x_{11})	0.167	0.074	0.167	2.261
12	Food intake of family(x_{12})	0.161	0.110	0.161	1.465
13	Calorie intake of family(x_{13})	0.184	0.081	0.184	2.281
14	Health status (x_{14})	0.224	0.081	0.224	2.745
15	Cost of media gadgets(x_{15})	0.033	0.086	0.033	0.378
16	Cost of energy consumption(x_{16})	-0.051	0.086	-0.051	-0.599
17	Monthly Cost of smart phone(x_{17})	0.167	0.081	0.167	2.053
18	Stress (x_{18})	-0.400	0.099	-0.400	-4.034

R square: 66.70%

The standard error of the estimate: 0.638

Results: Table 6.15 presents the full model of multiple regression analysis between 18 causal variables (x_1-x_{18}) and neighbours happiness (y_4).

Revelation: The table-6.15 presents multiple regression analysis (full model) between 18 causal variables (x_1-x_{18}) and neighbours happiness (y_3).The full Figure of multiple regression analysis portraits that with the combination of

18 causal variables, 66.70 percent variance in neighbours happiness (y_4), has been explained.

Stress (x_{18}) has got the highest beta value which is -0.400

Table 6.16. Stepwise Regression Analysis: Neighbour happiness (y_4) vs. 18 Causal Variables

Sl.No	Variables	Reg.coef. B	S.E. B	Beta	t value
1	Stress (x_{18})	-0.443	0.074	-0.443	-6.018
2	Calorie intake of family(x_{13})	0.248	0.065	0.248	3.793
3	Size of land holdings (x_3)	0.120	0.070	0.120	1.716
4	Monthly Cost of smart phone(x_{17})	0.188	0.067	0.188	2.817
5	Health status (x_{14})	0.198	0.076	0.198	2.610
6	Education (x_2)	0.182	0.066	0.182	2.768
7	Communication status (x_{11})	0.174	0.065	0.174	2.666

R square: 62.90%

The standard error of the estimate: 0.632

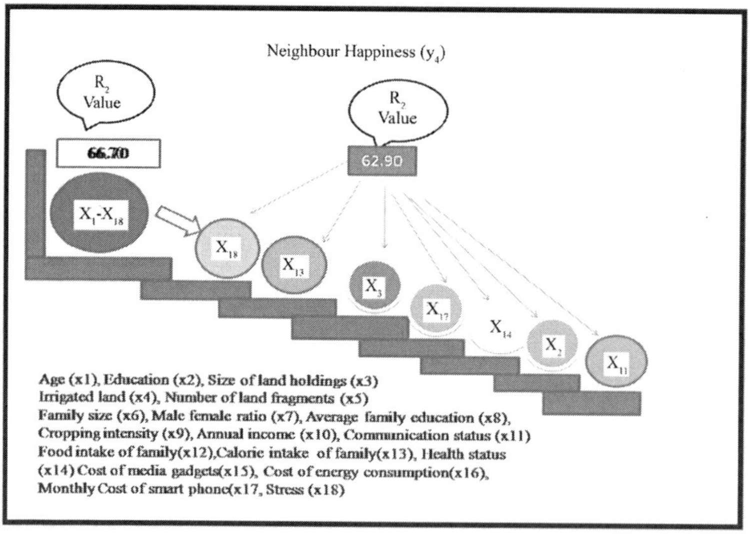

Fig. 6.11. Stepwise Regression Analysis: Neighbour happiness (y_4) vs. 18 Causal Variables

Results: We have to move for the next table (table 6.16) to isolate the causal dominant variables contributing to the level of neighbours happiness.

Revelation: It has been found that Stress (x_{18}), Calorie intake of family (x_{13}), Size of land holdings (x_3), Monthly Cost of smart phone (x_{17}), Health status (x_{14}), Education (x_2), Communication status (x_{11}), have been retained at the last step and has substantially contributed to the consequent variable neighbours happiness (y_4).

- The higher stress (x_{18}), lower will be neighbors' happiness (y_4).
- The higher calorie intake of family (x_{13}), higher will be neighbors happiness (y_4).
- The higher size of land holdings (x_3), higher will be neighbors' happiness (y_4).
- The higher monthly Cost of smart phone (x_{17}), higher will be neighbors' happiness (y_4).
- The higher health status (x_{14}), higher will be neighbors happiness (y_4).
- The higher education (x_2), higher will be neighbors happiness (y_4).
- The higher communication status (x_{11}), higher will be neighbors happiness (y_4).

These 7 variables together have contributed 62.90% of variance ie. $\left(\frac{62.90}{66.70}\times 100\right)$ = 94.30 per cent of total variance. The net contribution made by the 7 dominant causal variables stands at 94.30 percent. So these seven variables have got formidable causal impact.

Table 6.17. Path Analysis: Decomposition of Total Effect into Direct, Indirect and Residual Effect: Neighbours happiness y_4 vs. 18 exogenous variables

Sl.No	Variables	Total Effect	Direct Effect	Indirect Effect	Highest Indirect Effect
1	Age (x_1)	0.047	0.101	-0.054	-0.109 (x_2)
2	Education (x_2)	0.117	0.226	-0.109	-0.049 (x_1)
3	Size of land holdings (x_3)	0.347	0.138	0.209	0.087 (x_{14})
4	Irrigated land (x_4)	-0.061	-0.111	0.050	0.109 (x_5)
5	Number of land fragments (x_5)	0.014	0.136	-0.122	-0.09 (x_4)
6	Family size (x_6)	0.178	-0.068	0.246	0.129 (x_{18})
7	Male female ratio (x_7)	-0.100	0.104	-0.204	-0.055 (x_{18})
8	Average family education (x_8)	0.108	0.020	0.088	0.036 (x_3)
9	Cropping intensity (x_9)	-0.019	0.048	-0.067	-0.113 (x_{18})
10	Annual income (x_{10})	-0.030	0.016	-0.046	-0.064 (x_{13})
11	Communication status (x11)	0.198	0.168	0.030	0.044 (x_{18})
12	Food intake of family(x12)	0.491	0.159	0.332	0.232 (x_{18})
13	Calorie intake of family(x13)	0.356	0.185	0.171	0.067 (x_{18})
14	Health status (x14)	0.456	0.223	0.233	0.169 (x18)
15	Cost of media gadgets(x15)	-0.017	0.032	-0.049	0.047 (x5)

Sl.No	Variables	Total Effect	Direct Effect	Indirect Effect	Highest Indirect Effect
16	Cost of energy consumption(x16)	-0.084	-0.050	-0.034	0.041 (x_2)
17	Monthly Cost of smart phone (X_{17})	0.339	0.168	0.171	0.101 (x_{18})
18	Stress (X_{18})	-0.654	-0.400	-0.254	-0.094 (x_{14})

Residual effect: 0.332

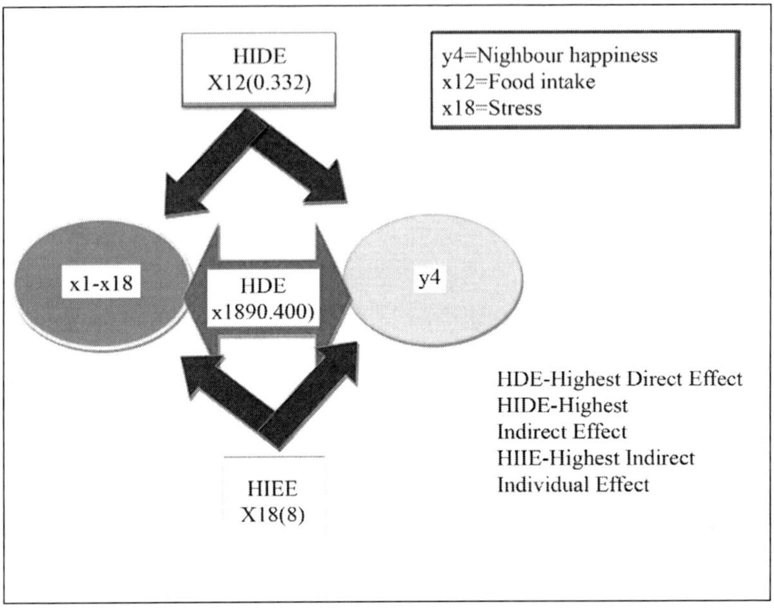

Fig. 6.12. Path Analysis: Decomposition of Total Effect into Direct, Indirect and Residual Effect: Neighbour happiness y_4 vs. 18 exogenous variables

Results: Table 6.17 presents the path analysis wherein the total effect of exogenous variable on consequent variable has been decomposed into direct, indirect and residual effect.

Revelation: It has been found that highest direct effect has been exerted by causal variable stress (X_{18}), and the highest indirect effect has been exerted by Family food intake (X_{12}). And, interestingly it has routed the highest indirect effect of 8 other causal variables to ultimately characterized neighbours happiness (Y_4).

Table 6.18. Coefficient of Correlation (r): Mental health (y_5) Vs. 18 Independent Variables

Sl.No.	Independent Variables	'r' Value	Remarks
1	Age (x_1)	-0.023	
2	Education (x_2)	-0.248	*
3	Size of land holdings (x_3)	-0.027	
4	Irrigated land (x4)	0.047	
5	Number of land fragments (x_5)	0.061	
6	Family size (x_6)	-0.150	
7	Male female ratio (x_7)	0.148	
8	Average family education (x_8)	-0.106	
9	Cropping intensity (x_9)	-0.120	
10	Annual income (x_{10})	-0.224	*
11	Communication status (x_{11})	-0.007	
12	Food intake of family(x_{12})	0.018	
13	Calorie intake of family(x_{13})	0.086	
14	Health status (x_{14})	0.044	
15	Cost of media gadgets(x^{15})	0.385	**
16	Cost of energy consumption (x^{16})	0.135	
17	Monthly Cost of smart phone(x_{17})	0.075	
18	Stress (x_{18})	0.003	

**Correlation is significant at the 0.01 level

*Correlation is significant at the 0.05 level

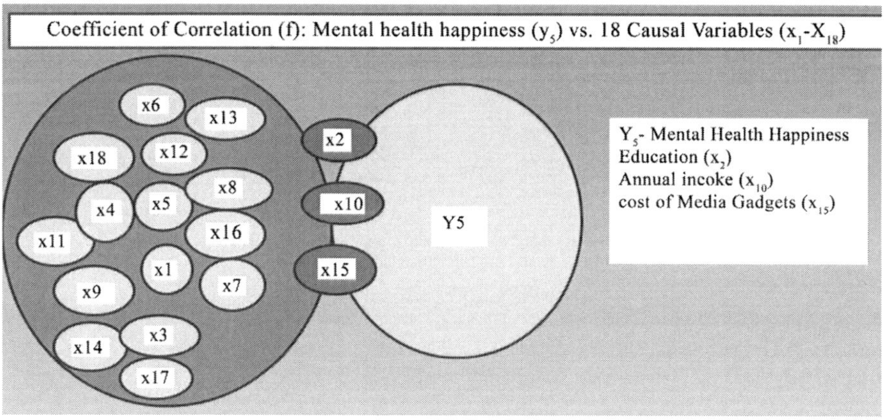

Fig. 6.13. Coefficient of Correlation (r): Mental health (y_5) Vs. 18 Independent Variables

Results: Table-6.18 presents the coefficient of correlation between **mental health happiness (y_5)** and 18 independent variables. it has been found that the following variables education (x_2)annual income (x_{10})cost of media gadgets (x_{15})have been recorded significant correlation with the dependent variable, mental health happiness (y_5).

Revelation: The higher is the **education (x_2),** the more would be the mind argumentation. The reasoning will go deeper and contemplation will go deeper with the up scaling of education. Stereotypes of happiness may be affected with the scaling up of education so mental health will decline.

The higher is the **income (x_{10})** the worst have been the mental health. Any money beyond what we need moves in a vicious way and invites unhappiness.

Exposure to electronics **media gadgets (x_{15})**to an optimum level balances and maintain psychomotor dynamics in such a way that it doesn't effect mental happiness and physical wellbeing. But beyond the optimum it would be disruptive.

Table 6.19. Multiple Regression Analysis: Mental health (y_5) Vs. 18 Causal Variables

Sl.No.	Variables	Reg. Coef. B	S.E. B	Beta	t Value
1	Age (x1)	-0.138	0.112	-0.138	-1.235
2	Education (x_2)	-0.314	0.110	-0.314	-2.856
3	Size of land holdings (x3)	0.140	0.112	0.140	1.250
4	Irrigated land (x4)	0.055	0.165	0.055	0.332
5	Number of land fragments (x5)	-0.078	0.159	-0.078	-0.490
6	Family size (x6)	-0.138	0.110	-0.138	-1.258
7	Male female ratio (x7)	0.206	0.099	0.206	2.075
8	Average family education (x8)	-0.022	0.097	-0.022	-0.222
9	Cropping intensity (x9)	-0.166	0.129	-0.166	-1.288
10	Annual income (x10)	-0.112	0.114	-0.112	-0.975
11	Communication status (x11)	0.087	0.103	0.087	0.845
12	Food intake of family(x12)	-0.038	0.153	-0.038	-0.248
13	Calorie intake of family(x13)	0.060	0.112	0.060	0.534
14	Health status (x14)	0.024	0.113	0.024	0.214
15	Cost of media gadgets(x15)	0.446	0.120	**0.446**	3.713
16	Cost of energy consumption(x16)	-0.049	0.119	-0.049	-0.412
17	Monthly Cost of smart phone(x17)	0.175	0.113	0.175	1.542
18	Stress (x18)	0.072	0.138	0.072	0.520

R square: 35.70%

The standard error of the estimate: 0.886

Results: Table-6.19 presents the full model of multiple regression analysis between 18 causal variables (x_1-x_{18}) and Mental health happiness (y_5).

Revelation: The table-6.19 presents multiple regression analysis (full Figure) between 18 causal variables (x_1-x_{18}) and Mental health happiness (y_5).The full Figure of multiple regression analysis portraits that with the combination of

18 causal variables, 35.70 percent variance in Mental health happiness (y_5)has been explained.

Cost of media gadgets (x_{15}) has got the highest beta value which is 0.446

Table 6.20. Stepwise Regression Analysis: Mental Health (y_5) Vs. 18 Causal Variables

Sl.No	Variables	Reg.coef. B	S.E. B	Beta	t value
1	Media gadgets cost(x15)	0.402	0.088	0.402	4.549
2	Education (x_2)	-0.298	0.089	-0.298	-3.350
3	Male female ratio (x7)	0.180	0.089	0.180	2.024

R square: 25.50%

The standard error of the estimate: 0.876

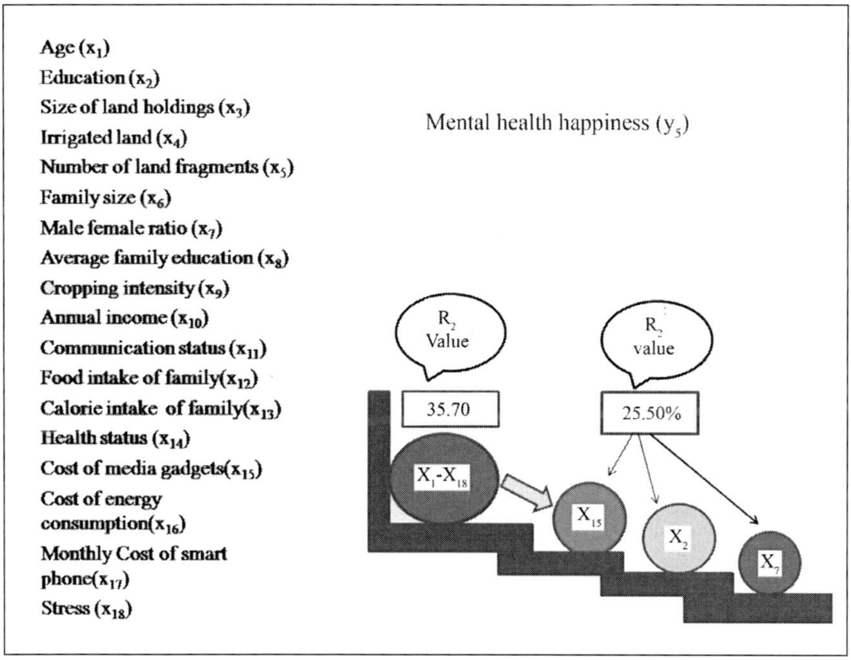

Fig. 6.14. Stepwise Regression Analysis: Mental health Happiness (y_5) Vs. 18 Causal Variables

Results: We have to move for the next table (table 6.20) to isolate the causal dominant variables contributing to the level of mental health happiness (y_5).

Revelation: It has been found that media gadgets cost(x_{15})education (x_2) male female ratio (x_7)have been retained at the last step and has substantially contributed to the consequent variable mental health happiness (y_5).

- The higher media gadgets cost(x_{15}) higher will be the mental health happiness (y_5).
- The higher education (x_2) lower will be the mental health happiness(y_5).
- The higher male female ratio (x_7) higher will be the mental health happiness(y_5).

These 3 variables together have contributed 62.90% of variance i.e. (= 71.42. The net contribution made by the 3 dominant causal variables stands at 71.42 percent. So these three variables have got formidable causal coefficient.

Table 6.21. Path Analysis: Decomposition of Total Effect into Direct, Indirect and Residual Effect: Mental Health Happiness y5 vs. 18 exogenous variables

Sl.No	Variables	Total Effect	Direct Effect	Indirect Effect	Highest Indirect Effect
1	Age (x1)	-0.023	-0.138	0.115	0.152 (x_2)
2	Education (x_2)	-0.248	-0.314	0.066	0.067 (x_1)
3	Size of land holdings (x3)	-0.027	0.139	-0.166	-0.081 (x_{15})
4	Irrigated land (x4)	0.047	0.052	-0.005	0.154 (x_{15})
5	Number of land fragments (x5)	0.061	-0.076	0.137	0.152 (x_{15})
6	Family size (x6)	-0.150	-0.138	-0.012	-0.030 (x_2)
7	Male female ratio (x7)	0.148	0.205	-0.057	-0.042 (x_2)
8	Average family education (x8)	-0.106	-0.021	-0.085	-0.044 (x_2)
9	Cropping intensity (x9)	-0.120	-0.166	0.046	0.046 (x_{17})
10	Annual income (x10)	-0.224	-0.111	-0.113	-0.060 (x_{15})
11	Communication status (x11)	-0.007	0.086	-0.093	0.055 (x_2)
12	Food intake of family(x12)	0.018	-0.038	0.056	0.064 (x_9)
13	Calorie intake of family(x13)	0.086	0.060	0.026	0.038 (x_{10})
14	Health status (x14)	0.044	0.024	0.020	0.055 (x_3)
15	Cost of media gadgets(x15)	0.385	0.446	-0.061	-0.026 (x_5)
16	Cost of energy consumption(x16)	0.135	-0.049	0.184	0.22 (x_{15})
17	Monthly Cost of smart phone(x17)	0.075	0.174	-0.099	-0.048 (x_7)
18	Stress (x18)	0.003	0.071	-0.068	-0.047 (x_9)

Residual effect: 0.643

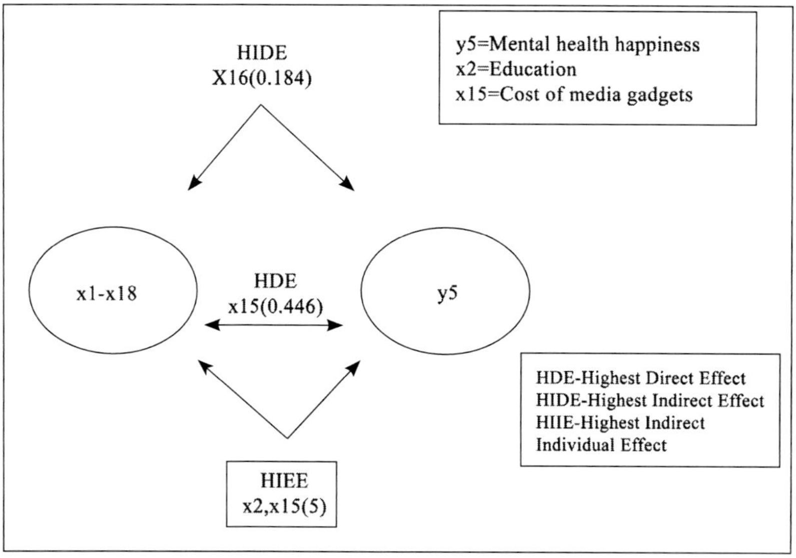

Fig. 6.15. Path Analysis: Decomposition of Total Effect into Direct, Indirect and Residual Effect: Mental Health Happiness (y_5) vs. 18 exogenous variables

Results: Table 6.21 presents the path analysis wherein the total effect of exogenous variable on consequent variable has been decomposed into direct, indirect and residual effect.

Revelation: It has been found that highest direct effect has been exerted by causal variable cost of media gadgets (x15) and the highest indirect effect has been exerted by energy consumption (x16). And interestingly it has routed the highest indirect effect of 5 other causal variables to ultimately characterized mental health happiness(y5).

Table 6.22. Coefficient of Correlation (r): Weather Related Happiness (y_6) Vs. 18 Independent Variables

Sl.No.	Independent Variables	'r' Value	Remarks
1	Age (x1)	0.027	
2	Education (x_2)	0.113	
3	Size of land holdings (x3)	0.112	
4	Irrigated land (x4)	-0.136	
5	Number of land fragments (x5)	-0.128	
6	Family size (x6)	-0.021	
7	Male female ratio (x7)	-0.062	

Sl.No.	Independent Variables	'r' Value	Remarks
8	Average family education (x8)	-0.189	
9	Cropping intensity (x9)	-0.310	**
10	Annual income (x10)	-0.148	
11	Communication status (x11)	-0.138	
12	Food intake of family(x12)	0.705	**
13	Calorie intake of family(x13)	0.134	
14	Health status (x14)	0.162	
15	Cost of media gadgets(x15)	-0.055	
16	Cost of energy consumption(x16)	0.168	
17	Monthly Cost of smart phone(x17)	0.368	**
18	Stress (x18)	-0.556	**

**Correlation is significant at the 0.01 level

*Correlation is significant at the 0.05 level

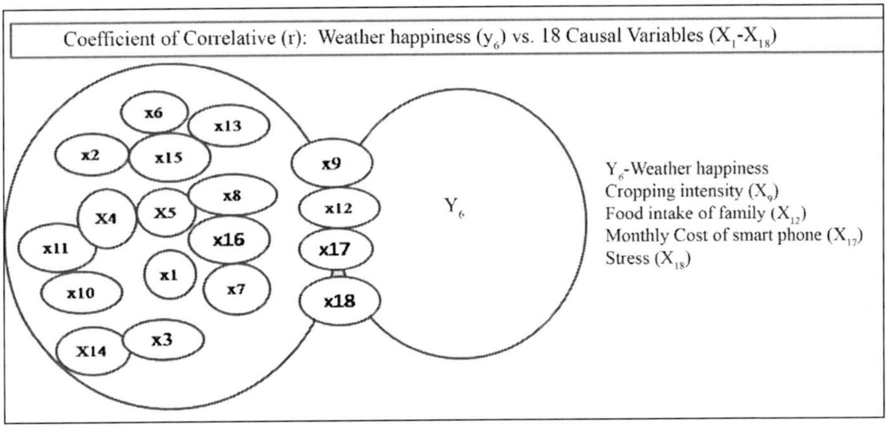

Fig. 6.16. Coefficient of Correlation (r): Weather Related Happiness (y_6) Vs. 18 Independent Variables

Results: Table-6.22 presents the coefficient of correlation between **weather happiness (y_6)** and 18 independent variables. it has been found that the following variables cropping intensity (x_9)food intake of family(x_{12})monthly cost of smart phone (x_{17})stress (x18)have recorded significant correlation with the dependent variable, weather happiness (y_6).

Revelation: Cropping intensity (x_9) when cropping intensity goes up, number of crops increases by default. Every crop has got unique weather requirements; some crops may need high intensity rainfall while others adopt to survive in a low water regime condition so with cropping intensity and with high yield potential the weather related uncertainty also escalates.

Food intake of family(x_{12}) as well as calorie intake can ensure better health and nutrition management and offers a psychomotor stability for running life with happiness and balances. Low calorie intake causes anemic disorders, fragile health all accounting to instability and disruption in the family and ecology. When weather goes better and supportive happiness splashes their winds of ecstasy.

Smart phone (x_{17}) helps access weather information from various sources, which ensures availability of weather forecasting appropriate crop management techniques and disaster related yeomen information.

Weather and psychic health are closely related. Prolonged drought incessant rainfall along with cloudy sky stormy nights and thundering sky all are negatively associated with **stress (x_{18})** and dereliction in body and mind.

Table 6.23. Multiple Regression Analysis: Weather happiness (y_6) vs. 18 Causal Variables

Sl.No.	Variables	Reg. Coef. B	S.E. B	Beta	t Value
1	Age (x1)	0.071	0.083	0.071	0.864
2	Education (x_2)	0.159	0.081	0.159	1.956
3	Size of land holdings (x3)	0.141	0.082	0.141	1.708
4	Irrigated land (x4)	-0.059	0.121	-0.059	-0.490
5	Number of land fragments (x5)	0.047	0.117	0.047	0.400
6	Family size (x6)	-0.088	0.081	-0.088	-1.089
7	Male female ratio (x7)	-0.002	0.073	-0.002	-0.030
8	Average family education (x8)	-0.151	0.072	-0.151	-2.099
9	Cropping intensity (x9)	-0.122	0.095	-0.122	-1.284
10	Annual income (x10)	0.029	0.084	0.029	0.345
11	Communication status (x11)	-0.100	0.076	-0.100	-1.315
12	Food intake of family(x12)	0.448	0.113	0.448	3.970
13	Calorie intake of family(x13)	-0.027	0.083	-0.027	-0.324
14	Health status (x14)	-0.073	0.084	-0.073	-0.877
15	Cost of media gadgets(x15)	-0.032	0.089	-0.032	-0.358
16	Cost of energy consumption(x16)	0.065	0.088	0.065	0.742
17	Monthly Cost of smart phone(x17)	0.191	0.084	0.191	2.290
18	Stress (x18)	-0.272	0.102	-0.272	-2.674

R square: 65.00%

The standard error of the estimate: 0.654

Results: Table-6.23 presents the full model of multiple regression analysis between 18 causal variables (x_1-x_{18}) and weather happiness (y_6).

Revelation: The table-6.23 presents multiple regression analysis (full Figure) between 18 causal variables (x_1-x_{18}) and weather happiness (y_6).The full Figure of multiple regression analysis portraits that with the combination of 18 causal variables, 65.00 percent variance in weather happiness (y_6)has been explained.

Table 6.24. Stepwise Regression Analysis: (y_6) Vs. 18 Causal Variables

Sl.No	Variables	Reg.coef. B	S.E. B	Beta	t value
1	Food intake of family(x12)	0.542	0.081	0.542	6.657
2	Monthly Cost of smart phone(x17)	0.170	0.069	0.170	2.468
3	Stress (x18)	-0.218	0.082	-0.218	-2.654
4	Communication status (x11)	-0.166	0.066	-0.166	-2.500

R square: 58.90%

The standard error of the estimate: 0.655

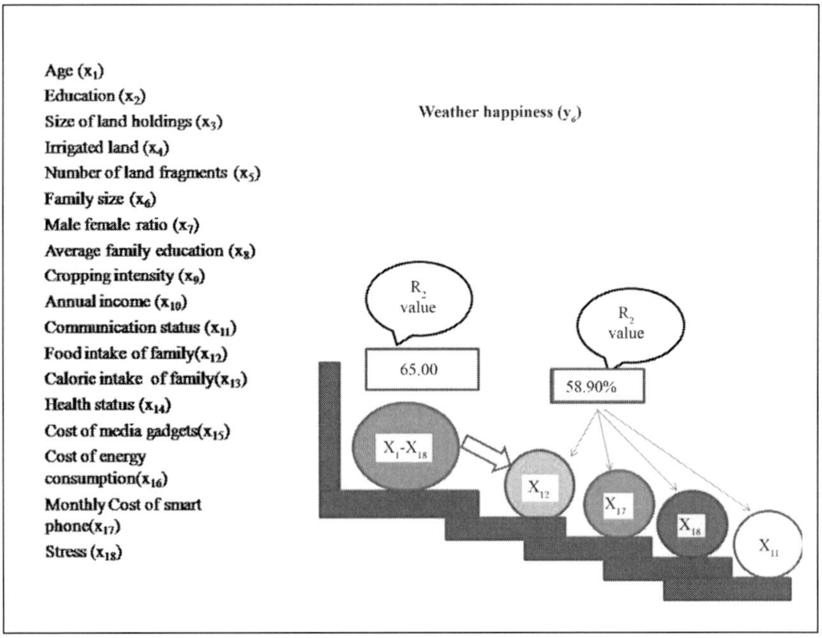

Fig. 6.17. Stepwise Regression Analysis: weather happiness (y_6) Vs. 18 Causal Variables

Results: We have to move for the next table (table 6.23) to isolate the causal dominant variables contributing to the level of weather happiness (y_6).

Revelation: It has been found that food intake of family(x_{12})monthly cost of smart phone(x_{17}) stress (x_{18}) communication status (x_{11}) have been retained at the last step and has substantially contributed to the consequent variable weather happiness (y_6).

- The higher food intake of family(x12) higher will be weather happiness (y_6).
- The higher monthly Cost of smart phone(x17) higher will be weather happiness (y_6).
- The higher stress (x18) lower will be weather happiness (y_6).
- The higher communication status (x11) lower will be weather happiness (y_6).

These 4 variables together have contributed 62.90% of variance ie. $\left(\frac{58.90}{65.00}\times 100\right)$ 90.61 per cent of total variance. The net contribution made by the 4 dominant causal variables stands at 90.61 percent. So these four variables have got formidable causal contribution.

Table 6.25. Path Analysis: Decomposition of Total Effect into Direct, Indirect and Residual Effect: weather happiness y6 vs. 18 exogenous variables

Sl.No	Variables	Total Effect	Direct Effect	Indirect Effect	Highest Indirect Effect
1	Age (x1)	0.027	0.072	-0.045	-0.077 (x_2)
2	Education (x_2)	0.113	0.159	-0.046	-0.035 (x1)
3	Size of land holdings (x3)	0.112	0.141	-0.029	0.05 (x18)
4	Irrigated land (x4)	-0.136	-0.058	-0.078	-0.071 (x12)
5	Number of land fragments (x5)	-0.128	0.045	-0.173	-0.07 (x12)
6	Family size (x6)	-0.021	-0.089	0.068	0.088 (x18)
7	Male female ratio (x7)	-0.062	-0.003	-0.059	-0.045 (x17)
8	Average family education (x8)	-0.189	-0.151	-0.038	-0.055 (**x12**)
9	Cropping intensity (x9)	-0.310	-0.123	-0.187	-0.17 (**x12**)
10	Annual income (x10)	-0.148	0.030	-0.178	-0.144 (**x12**)
11	Communication status (x11)	-0.138	-0.100	-0.038	0.03 (x18)
12	Food intake of family(x12)	**0.705**	**0.446**	0.259	0.158 (x18)
13	Calorie intake of family(x13)	0.134	-0.025	0.159	0.146 (**x12**)
14	Health status (x14)	0.162	-0.074	0.236	0.115 (x18)
15	Cost of media gadgets(x15)	-0.055	-0.032	-0.023	-0.062 (**x12**)
16	Cost of energy consumption(x16)	0.168	0.066	0.102	0.066 (**x12**)
17	Monthly Cost of smart phone(x17)	0.368	0.192	0.176	0.111 (**x12**)
18	Stress (x18)	-0.556	-0.272	-0.284	-0.258 (**x12**)

Residual effect: 0.350

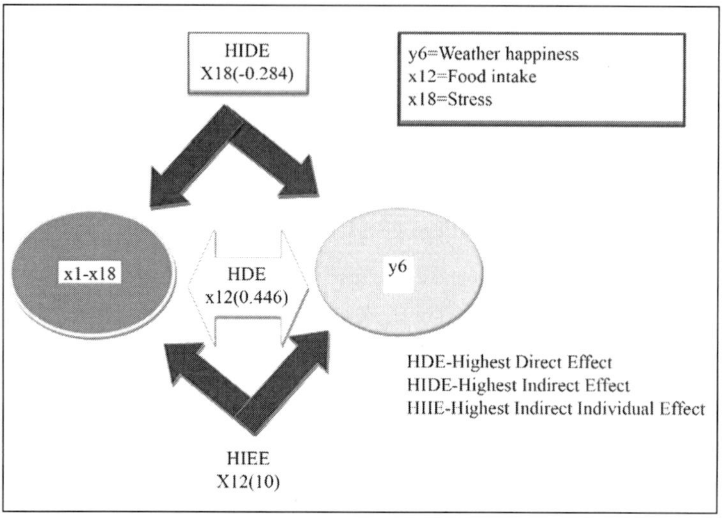

Fig. 6.18. Path Analysis: Decomposition of Total Effect into Direct, Indirect and Residual Effect: weather happiness y6 vs. 18 exogenous variables

Results: Table 6.25 presents the path analysis wherein the total effect of exogenous variable on consequent variable has been decomposed into direct, indirect and residual effect.

Revelation: It has been found that highest direct effect has been exerted by causal variable, food intake (x12) and the highest indirect effect has been exerted by stress (x18). And interestingly it has routed the highest indirect effect of 10 other causal variables to ultimately characterized weather happiness (y6).

Table 6.26. Coefficient of Correlation (r): Ecosytem Happiness (y_7) Vs. 18 Independent Variables

Sl.No.	Independent Variables	'r' Value	Remarks
1	Age (x1)	-0.052	
2	Education (x_2)	0.025	
3	Size of land holdings (x3)	0.173	
4	Irrigated land (x4)	-0.143	
5	Number of land fragments (x5)	-0.170	
6	Family size (x6)	0.044	
7	Male female ratio (x7)	-0.168	
8	Average family education (x8)	0.068	
9	Cropping intensity (x9)	0.496	**
10	Annual income (x10)	0.085	
11	Communication status (x11)	0.052	
12	Food intake of family(x12)	0.069	
13	Calorie intake of family(x13)	0.201	*

Sl.No.	Independent Variables	'r' Value	Remarks
14	Health status (x14)	0.118	
15	Cost of media gadgets(x15)	-0.235	*
16	Cost of energy consumption(x16)	-0.234	*
17	Monthly Cost of smart phone(x17)	0.470	**
18	Stress (x18)	0.055	

**Correlation is significant at the 0.01 level
*Correlation is significant at the 0.05 level

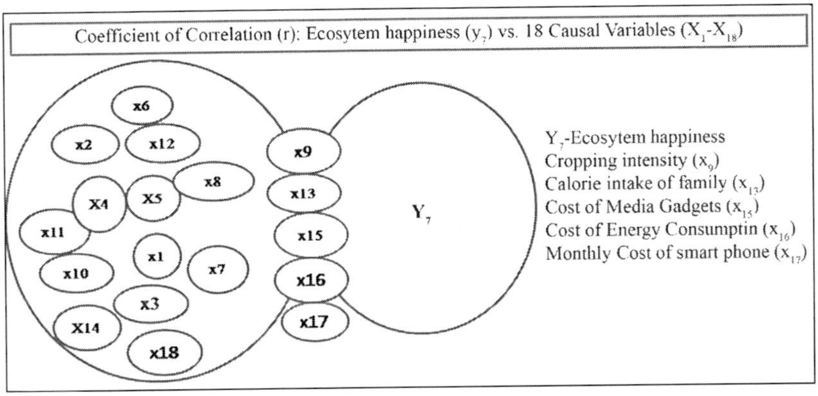

Fig. 6.19. Coefficient of Correlation (r): Ecosystem happiness (y_7) Vs. 18 Independent Variables

Results: Table 6.26 presents the coefficient of correlation between **ecosystem happiness (y_7)** and 18 independent variables. it has been found that the following variables cropping intensity (x_9)calorie intake of family (x_{13})cost of media gadgets (x_{15})cost of energy consumption (x_{16})monthly cost of smart phone (x_{17}) have recorded significant correlation with the dependent variable, ecosystem happiness (y_7).

Revelation: Cropping intensity (x_9) indicates diverse and secure crop yield and yield related farm income. So, with higher cropping intensity ecology happiness has been escalated

Food intake of family as well as **calorie intake (x_{13})** can ensure better health and nutrition management and offers a psychomotor stability for running life with happiness and balances. Low calorie intake causes anemic disorders, fragile health all accounting to instability and disruption in the family and ecology. When weather goes better and supportive happiness splashes their winds of ecstasy.

The higher has been the **media gadgets and energy consumption** cost the crux of sociological conflict goes up. It has a negative correlation with the ecological happiness. The farmers are using vigorous amount of fertilizer effects their farm eco system and yield will be stagnant.

Smart phone (x_{17}) helps access weather information from various sources, which ensures availability of weather forecasting appropriate crop management techniques and disaster related yeomen information

Table 6.27. Multiple Regression Analysis: Ecosystem happiness (y_7) Vs. 18 Causal Variables

Sl.No.	Variables	Reg. Coef. B	S.E. B	Beta	t Value
1	Age (x1)	-0.140	0.093	-0.140	-1.497
2	Education (x_2)	-0.060	0.092	-0.060	-0.655
3	Size of land holdings (x3)	-0.145	0.093	-0.145	-1.554
4	Irrigated land (x4)	0.191	0.137	0.191	1.388
5	Number of land fragments (x5)	-0.173	0.133	-0.173	-1.302
6	Family size (x6)	0.139	0.092	0.139	1.519
7	Male female ratio (x7)	-0.011	0.083	-0.011	-0.136
8	Average family education (x8)	-0.004	0.081	-0.004	-0.049
9	Cropping intensity (x9)	0.479	0.108	0.479	4.448
10	Annual income (x10)	0.067	0.096	0.067	0.702
11	Communication status (x11)	-0.020	0.086	-0.020	-0.237
12	Food intake of family(x12)	0.283	0.128	0.283	2.217
13	Calorie intake of family(x13)	0.144	0.094	0.144	1.531
14	Health status (x14)	0.265	0.095	0.265	2.802
15	Cost of media gadgets(x15)	-0.182	0.100	-0.182	-1.817
16	Cost of energy consumption(x16)	-0.056	0.100	-0.056	-0.566
17	Monthly Cost of smart phone(x17)	0.353	0.095	0.353	3.731
18	Stress (x18)	0.310	0.115	0.310	2.696

R square: 55.20%

The standard error of the estimate: 0.740

Results: Table-6.27 presents the full model of multiple regression analysis between 18 causal variables (x_1-x_{18}) and ecosystem happiness (y_7).

Revelation: The table-6.27 presents multiple regression analysis (full figure) between 18 causal variables (x_1-x_{18}) and ecosystem happiness (y_7).the full figure of multiple regression analysis portraits that with the combination of 18 causal variables, 55.20 percent variance in ecosystem happiness (y_7)has been explained.

Table 6.28. Stepwise Regression Analysis: Ecosystem happiness (y_7) Vs. 18 Causal Variables

Sl.No	Variables	Reg.coef. B	S.E. B	Beta	t value
1	Cropping intensity (x9)	0.437	0.080	0.437	5.436
2	Monthly Cost of smart phone(x17)	0.346	0.079	0.346	4.368
3	Media gadgets cost(x15)	-0.237	0.076	-0.237	-3.130
4	Health status (x14)	0.175	0.078	0.175	2.251

R square: 45.60%

The standard error of the estimate: 0.753

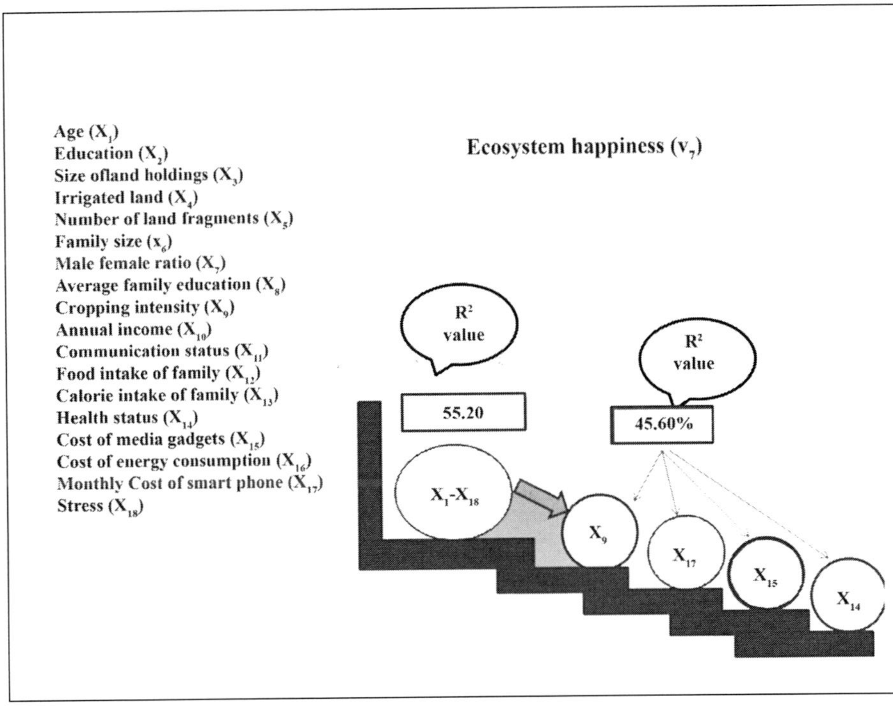

Fig. 6.20. Stepwise Regression Analysis: Ecosystem happiness (y_7) Vs. 18 Causal Variables

Results: We have to move for the next table (table 6.28) to isolate the causal dominant variables contributing to the level of ecosystem happiness (y_7).

Revelation: It has been found that Cropping intensity (x_9)Monthly Cost of smart phone(x_{17})Media gadgets cost(x_{15})Health status (x_{14}) have been retained at the last step and has substantially contributed to the consequent variable ecosystem happiness (y_7).

- The higher Cropping intensity (x9) higher has been ecosystem happiness (y_7).
- The higher Monthly Cost of smart phone(x17) higher has been ecosystem happiness (y_7).
- The higher Media gadgets cost(x15) lower has been ecosystem happiness (y_7).
- The higher Health status (x14) higher has been ecosystem happiness (y_7).

These 4 variables together have contributed 62.90% of variance i.e. $\left(\frac{45.60}{55.20}\times 100\right)$ = 82.60. The net contribution made by the 4 dominant causal variables stands at 82.60 percent. So the 4 variables have got formidable causal coefficient.

Table 6.29. Path Analysis: Decomposition of Total Effect into Direct, Indirect and Residual Effect: Ecosystem Happiness y_7 vs. 18 exogenous variables

Sl.No	Variables	Total Effect	Direct Effect	Indirect Effect	Highest Indirect Effect
1	Age (x1)	-0.052	-0.140	0.088	0.043 (x12)
2	Education (x_2)	0.025	-0.060	0.085	0.068 (x1)
3	Size of land holdings (x3)	0.173	-0.146	0.319	0.104 (x14)
4	Irrigated land (x4)	-0.143	0.192	-0.335	-0.14 (x5)
5	Number of land fragments (x5)	-0.170	-0.174	0.004	0.154 (x4)
6	Family size (x6)	0.044	0.140	-0.096	-0.101 (x18)
7	Male female ratio (x7)	-0.168	-0.012	-0.156	-0.081 (x17)
8	Average family education (x8)	0.068	-0.004	0.072	0.063 (x9)
9	Cropping intensity (x9)	0.496	0.480	0.016	-0.109 (x12)
10	Annual income (x10)	0.085	0.067	0.018	0.1 (x9)
11	Communication status (x11)	0.052	-0.020	0.072	0.094 (x9)
12	Food intake of family(x12)	0.069	0.284	-0.215	-0.184 (x9)
13	Calorie intake of family(x13)	0.201	0.143	0.058	0.094 (x12)
14	Health status (x14)	0.118	0.266	-0.148	-0.131 (x18)
15	Cost of media gadgets(x15)	-0.235	-0.183	-0.052	0.067 (x4)
16	Cost of energy consumption(x16)	-0.234	-0.056	-0.178	-0.118 (x9)
17	Monthly Cost of smart phone(x17)	0.470	0.352	0.118	0.125 (x9)
18	Stress (x18)	0.055	0.311	-0.256	-0.165 (x12)

Residual effect: 0.4481

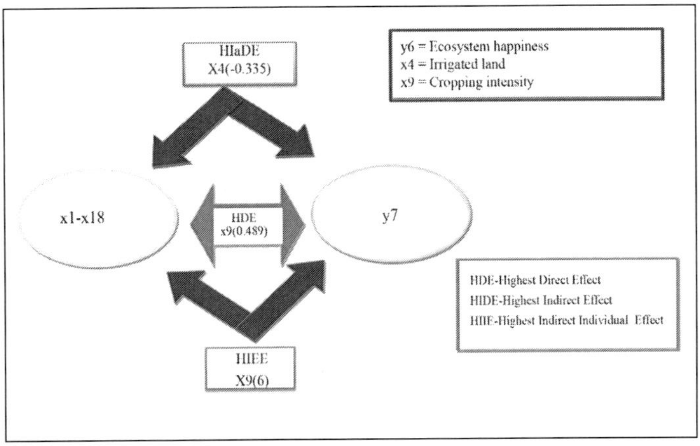

Fig. 6.21. Path Analysis: Decomposition of Total Effect into Direct, Indirect and Residual Effect: Ecosystem happiness y7 vs. 18 exogenous variables

Results: Table 6.29 presents the path analysis wherein the total effect of exogenous variable on consequent variable has been decomposed into direct, indirect and residual effect.

Revelation: It has been found that highest direct effect has been exerted by causal cropping intensity (x_9) ,and the highest indirect effect has been exerted by irrigated land (x_4). And interestingly it has routed the highest indirect effect of 6 other causal variables to ultimately characterized ecosystem happiness (y_7).

Table 6.30. Canonical Covariate Analysis: Interaction between Left Side Dependent and Right Side Independent Variables

Dependent Variable	Coefficients
Economic happiness (y_1)	-2.517
Productivity happiness (y_2)	3.412
Family happiness (y_3)	-0.547
Neighbours happiness (y_4)	-0.114
Mental health happiness (y_5)	0.190
Weather happiness(y_6)	-0.478
Ecosystem happiness (y7)	-0.023

Independent Variable	Coefficients
Age (x_1)	-0.063
Education (x_2)	0.073
Size of land holdings (x_3)	0.268
Irrigated land (x_4)	0.127

Number of land fragments (x5)	-0.074
Family size (x6)	-0.246
Male female ratio (x7)	0.169
Average family education (x8)	-0.080
Cropping intensity (x9)	0.227
Annual income (x10)	0.086
Communication status (x11)	-0.068
Food intake of family(x12)	0.397
Calorie intake of family(x13)	-0.537
Health status (x14)	-0.352
Cost of media gadgets(x15)	0.035
Cost of energy consumption(x16)	0.088
Monthly Cost of smart phone(x17)	-0.477
Stress (x18)	-0.582

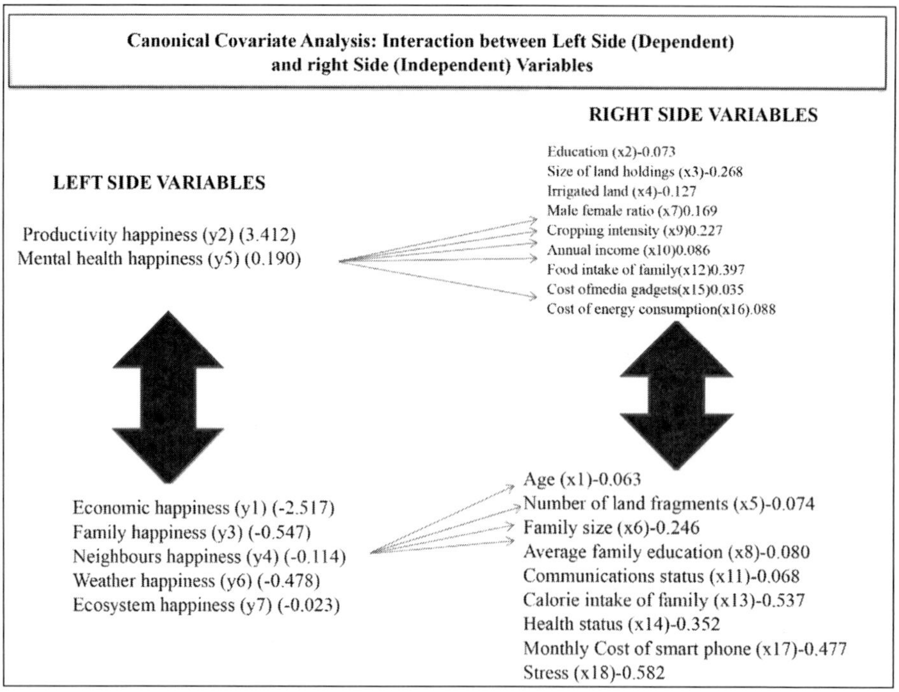

Figure 6.22. Canonical Covariate Analyses: Interaction between Left Side Dependent and Right Side Independent Variables

Revelation: Canonical covariant analysis helps us understand and visualize the nature of sub conglomeration between x and y variables. In this sub conglomeration the co-integrations of x and y variables becomes vivid and conspicuous. From the above illustrations it has been evince that productivity happiness **(y2)**mental health happiness **(y5)** prefers to move together so it is well discernable that, productivity happiness has developed intimate kinship with mental health happiness. It is well understood that, mental health or mental well being is expected to be supported by crop productivity viscerally crop related income happiness. Farmers become happy only when his lands are coming up expected productive behavior and security. This will ensure happiness for the farmers. Farmers are under challenges and stress from the unpredictable behaviour. When these two leftside consequent variables are moving together, in one direction. The right side variables education (x_2), size of land holdings ($\mathbf{x_3}$), irrigated land ($\mathbf{x_4}$), male female ratio ($\mathbf{x_7}$), cropping intensity **($\mathbf{x_9}$),** annual income ($\mathbf{x_{10}}$)food intake of family($\mathbf{x_{12}}$)cost of media gadgets($\mathbf{x_{15}}$)cost of energy consumption($\mathbf{x_{16}}$) selected and classified based on cross loading, are organically linked with the left side conglomeration of y variables.

Similarly, the following y variables have created the sub-conglomeration viz. economic happiness (y_1), family happiness (y_3), neighbours happiness (y_4), weather happiness (y_6), ecosystem happiness (y_7). The probable reasons why these y variables do form a sub conglomeration together are that both the family and neighbours are thriving in a close ecosystem. The proximity of crop land with the homestead land promotes better and diverse involvement of family labour into all kinds of agricultural operation. Even in inclement weather the involvement of family labour into all kind of agricultural operation. Even in inclement weather the involvement of family labour with possible if it is proximately located with the homestead land that is how and why all these five variables have formed sub conglomeration of y variables. again, when this constellation of variables is interacting together, thus the following right side variables age **($\mathbf{x_1}$)** number of land fragments **($\mathbf{x_5}$),** family size **($\mathbf{x_6}$),** average family education ($\mathbf{x_8}$), communication status **($\mathbf{x_{11}}$),** calorie intake of family($\mathbf{x_{13}}$), health status ($\mathbf{x_{14}}$), monthly cost of smart phone($\mathbf{x_{17}}$), stress ($\mathbf{x_{18}}$) in a clandestine way have built up a close proximity with the right variables.

Table 6.31. Factor Analyses: Strategic Conglomeration of Variables into Factor

Factors	Variables	Factor loading	% of variance	Cumulative %	Factor renamed
1	Size of land holdings (x_3) Health status (x_{14}) Monthly Cost of smart phone(x_{17}) Economic happiness (y_1) Productivity happiness (y_2) Family happiness (y_3) Neighbours happiness (y_4) Ecosystem happiness (y_7)	0.543 0.369 0.521 0.906 0.913 0.903 0.763 0.547	22.172	22.172	*Happiness ecology*
2	Cropping intensity (x_9) Food intake of family(x_{12}) Cost of energy consumption (x_{16}) Stress (x_{18}) Weather happiness(y_6)	-0.700 0.728 0.416 -0.617 0.681	12.287	34.460	*Social ecology with happiness*
3	Irrigated land (x_4) Fragments of land (x_5) Family size (x_6) Average family education (x_8) Media gadgets cost (x_{15})	0.813 0.790 0.466 0.338 0.487	9.655	44.114	*Resource networking*
4	Age (x_1) Education (x_2)	-0.669 0.757	7.520	51.635	*Personal profile*
5	Income per year(x_{10}) Calorie intake of family(x_{13})	-0.516 0.488	6.505	58.140	*Food economics*
6	Male female ratio (x_7)	-0.533	5.574	63.713	
7	Mental health happiness (y_5)	0.533	4.894	68.607	
8	Communication status (x_{11})	-0.608	4.091	72.698	

Results: Table- 6.31 presents the strategic conglomeration of variables into factor through factor analysis.

Revelation: Table 6.31 presents the principle component analysis .But the strategic conglomeration of operationally similar variables, called principle components based on factor loading. Factor analysis (Principle Component Analysis) is a data reduction approach. Here the total number of interacting variable, based on factor loading and Eigen values, has been grouped into identifiable factor. Each factor is retaining the homogeneous variable, strongly bound to each other towards taking a strategic decision for effective management of resource or organization

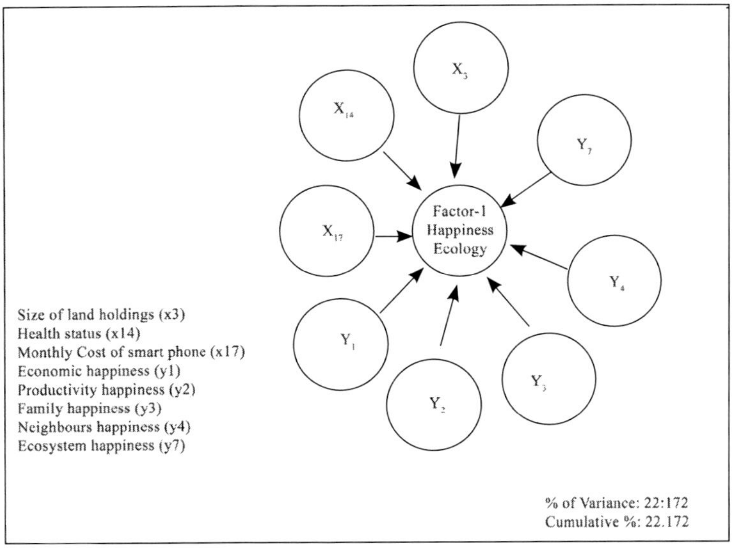

Fig. 6.23. Factor Analysis: 1 Happiness ecology

Factor 1 has Accommodated following ; size of land holdings (x_3) health status (x_{14})monthly cost of smart phone(x_{17})economic happiness (y_1)productivity happiness (y_2)family happiness (y_3)neighbours happiness (y_4)ecosystem happiness (y_7) has contributed 22.17 percent of variance. This factor has been renamed as ***happiness ecology.***

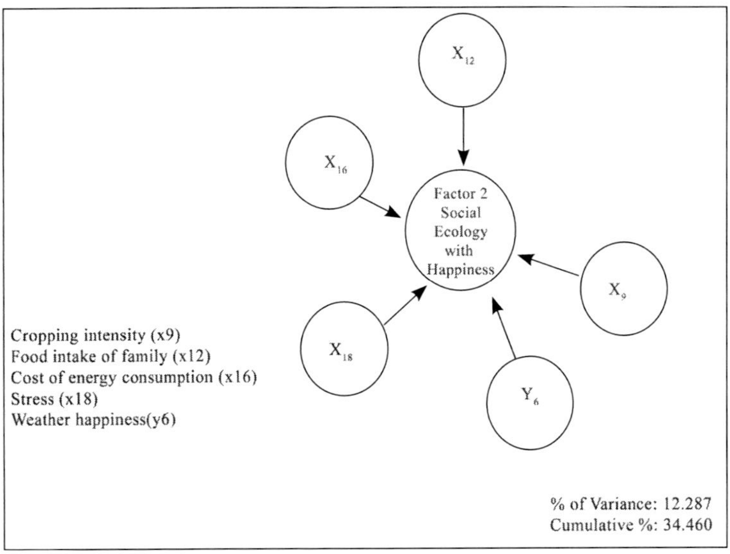

Fig. 6.24. Factor Analysis: 2 Social ecology with happiness

Factor 2 has Accommodated following cropping intensity (x_9), food intake of family(x_{12}),cost of energy consumption (x_{16}),stress (x_{18}) , happiness(y6); has contributed 12.287 percent of variance. This factor has been renamed as ***social ecology with happiness.***

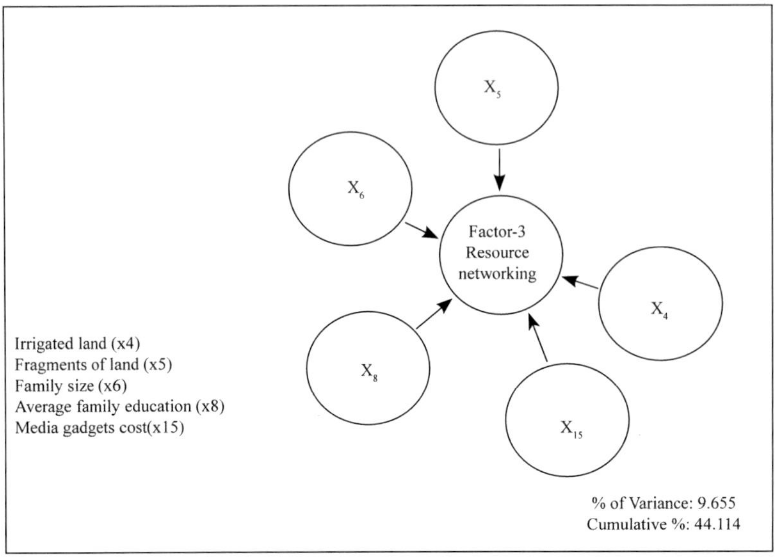

Fig. 6.25. Factor Analysis: 3 Resource networking

Factor 3 has accommodated following variables; irrigated land (x_4), fragments of land (x_5), family size (x_6),average family education (x_8),media gadgets cost(x_{15}) and, has contributed 9.655 percent of variance. This factor has been renamed as ***resource networking.***

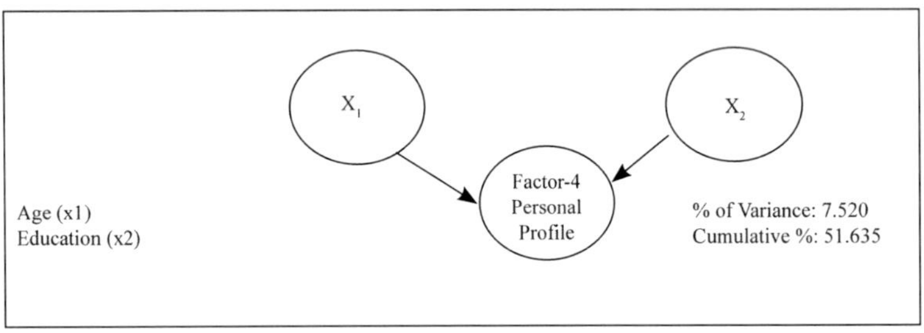

Fig. 6.26. Factor Analysis: 4 Personal profile

Factor 4 has accommodated following Age (x_1) Education (x_2)and , has contributed 7.520 percent of variance. This factor has been renamed as ***Personal profile.***

Figure 6.27 Factor Analysis: 5 Food economics

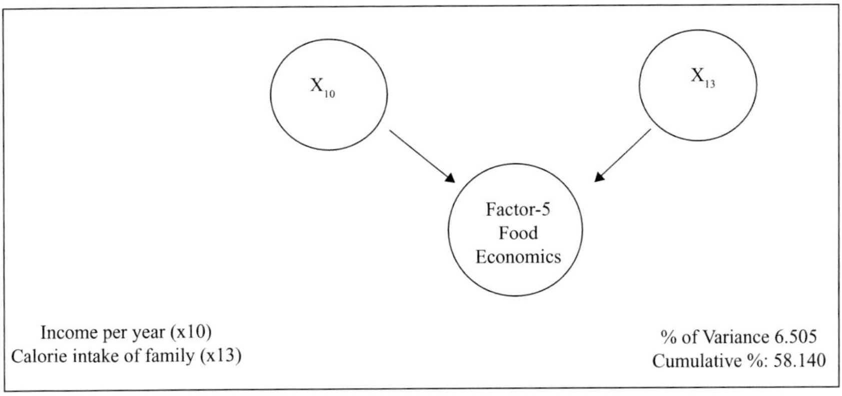

Fig. 6.27. Factor Analysis: 5 Food economics

Factor 5: Accommodation of following Income per year(x_{10}) calorie intake of family($\mathbf{x_{13}}$) has contributed 6.505 percent of variance. This factor has been renamed as ***Food economics.***

Factor 6: The variable male female ratio (x_7) is a solitary variable/factor. So , renaming is not necessary.

Factor 7: The variable mental health happiness (y_5) is a solitary factor. So, renaming of is not necessary.

Factor 8: The variable communication status (x_{11}) is a solitary factor. So, renaming is not necessary.

Artificial Neural Networking

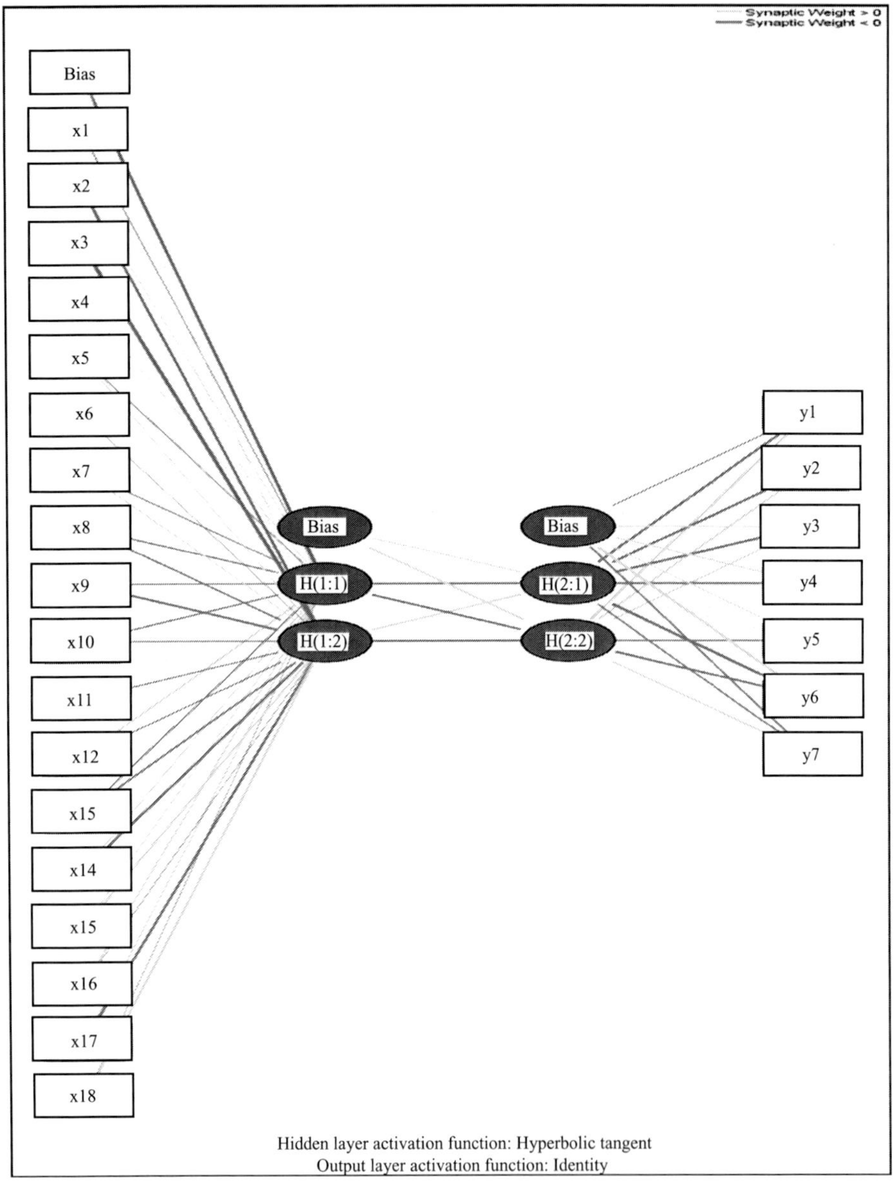

Fig. 6.28. Artificial Neural Networking

Figure Summary

Training	Sum of Squares Error		91.824
	Average Overall Relative Error		0.380
	Relative Error for Scale Dependents	Economic happiness (y1)	0.089
		Productivity happiness (y2)	0.162
		Family happiness (y3)	0.163
		Neighbours happiness (y4)	0.364
		Mental health happiness (y5)	0.872
		Weather happiness(y6)	0.334
		Ecosystem happiness (y7)	0.677
Testing	Sum of Squares Error		50.977
	Average Overall Relative Error		0.509
	Relative Error for Scale Dependents	Economic happiness (y1)	0.209
		Productivity happiness (y2)	0.326
		Family happiness (y3)	0.312
		Neighbours happiness (y4)	0.468
		Mental health happiness (y5)	1.033
		Weather happiness(y6)	0.409
		Ecosystem happiness (y7)	0.700

Artificial Neural Network has been carried out to measure the efficacy of input variables (x_1-x_{18}) , passing through the hidden layers and ultimately impacting on the output variables (y_{1}-y_7).

Number of hidden layers depends on the amount of errors as estimated for the training of input data. Hidden layers are made functional by adding bias to it to minimize the error. Now, the following variables have exerted the dominant impact on the output variables; education (x_2)size of land holdings (x_3),cropping intensity (x_9),calorie intake of family (x_{13}) health status (x_{14}),monthly cost of smart phone (x_{17}).

So, these input variables have selected by applying ANN are of immense strategic implications and these variables merit special attention to deal with happiness aspects of social ecology of under study.

7

Summary and Conclusion

The present study has been conducted in one of the agro-ecological zones of the state of West Bengal, New alluvial zone. Four villages from Haringhata block in Nadia district were selected from the new alluvial agro- ecological zone to conduct the study. One hundred farmers were selected from the villages of Dalilpur, Mitrapur, Fatehpur, 25 respondents from each. The primary data were collected with the help of structured interview schedule by following the personal interview method. The secondary data were collected from our departmental library, internet, Department of Agriculture, West Bengal, Bidhan Chandra krishi Viswavidyalaya, etc. for establishing the conceptual framework of the study.

The general objective of the study is to estimate and analyze the **"The Happiness of Farmers: The Socio-ecological Interpretation and Strategy Implications"** and the **specific objectives** are:

i. To conduct benchmark survey on overall happiness of respondents' terms and condition assigned in the study.

ii. To isolate and identify the variables impacting on happiness of farmers, the predicted character of the study.

iii. To elucidate the level direction and network of interaction amongst and between the set of predictor variables.

iv. Based on the empirical research to extract the policy implications both for micro and macro sociological aspects.

7.1 Summary

18 Independent variables and 7 Dependent variables were selected for the study respectively and measured with the help of exact scales developed by previous social science researchers or by modifying the developed scale by structured interview schedule for requirement of the study.

Independent variables selected for the study are, age (x_1) education (x_2) size of land holdings (x_3) irrigated land (x_4) fragments of land (x_5) family size (x_6) male female ratio (x_7) average family education (x_8) cropping intensity (x_9) annual income (x_{10}) communication status (x_{11}) family food intake (x_{12}) calorie intake (x_{13}) health status (x_{14}) media gadgets (x_{15}) cost of energy consumption (x_{16}) monthly cost of smart phone (x_{17}) stress (x_{18}) the dependent variables selected for the study are, financial happiness (y_1)productivity related happiness (y_2) family related happiness(y_3)neighbors related happiness (y_4) mental health related happiness (y_5)weather related happiness (y_6) ecology related happiness(y_7).

The statistical tools viz., Mean, Median, Mode, Standard Deviation, and Coefficient of variation, Correlation coefficient, Step wise Regression Analysis, Path analysis and Factor Analysis, ANN ,CCA were used for the purpose of the study.

Coefficient of Correlation (r): Economic happiness (y_1) Vs. 18 Independent Variables

Presents the coefficient of correlation between **Economic happiness (y_1)** and 18 independent variables. it has been found that the following variables education (x_2)size of land holdings (x_3),family size (x_6),average family education (x_8) cropping intensity (x_9)annual income (x_{10})communication status (x_{11})calorie intake of family (x_{13})cost of media gadgets (x_{15})cost of energy consumption (x_{16})monthly cost of smart phone (x_{17})stress (x_{18})have been recorded significant correlation with the dependent variable, economic happiness (y_1).

Coefficient of Correlation (r): Productivity happiness (y_2) Vs. 18 Independent Variables

It has been found that the following variables Size of land holdings (x_3) male female ratio (x_7)average family education (x_8),cropping intensity (x_9),communication status (x_{11}),food intake of family(x_{12}),calorie intake of family (x_{13}),health status (x_{14}), cost of media gadgets (x_{15}),cost of energy consumption (x_{16}),monthly cost of smart phone (x_{17}),stress (x_{18}) have been recorded significant correlation with the dependent variable, productivity happiness (y_2).

Coefficient of Correlation (r): Family happiness (y_3) Vs. 18 Independent Variables

It has been found that the following variables education ($_{x2}$)size of land holdings (x_3)family size (x_6)male female ratio (x_7) average family education (x_8)cropping intensity (x_9)communication status (x_{11})food intake of family(x_{12}) calorie intake of family (x_{13})

Monthly Cost of smart phone (x_{17})stress (x_{18})have been recorded significant correlation with the dependent variable, family happiness (y_3).

Coefficient of Correlation (r): Neighbours happiness (y_4) Vs. 18 Independent Variables

It has been found that the following variables size of land holdings (x_3) communication status (x_{11})food intake of family(x_{12})calorie intake of family (x_{13})health status (x_{14})monthly cost of smart phone (x_{17})stress (x_{18})have been recorded significant correlation with the dependent variable, neighbours happiness (y_4).

Coefficient of Correlation (r): Mental health happiness (y_5) Vs. 18 Independent Variables

It has been found that the following variables education ($_{x2}$) annual income (x_{10})cost of media gadgets (x_{15})have been recorded significant correlation with the dependent variable, mental health happiness (y_5).

Coefficient of Correlation (r): Weather happiness (y_6) Vs. 18 Independent Variable

It has been found that the following variables cropping intensity (x_9)food intake of family(x_{12})monthly cost of smart phone (x_{17})stress (x18)have been recorded significant correlation with the dependent variable, weather happiness (y_6).

Coefficient of Correlation (r): Ecosystem happiness (y_7) Vs. 18 Independent Variable

It has been found that the following variables cropping intensity (x_9)calorie intake of family (x_{13})cost of media gadgets (x_{15})cost of energy consumption (x_{16})monthly cost of smart phone (x_{17})have been recorded significant correlation with the dependent variable, ecosystem happiness (y_7).

Stepwise Regression Analysis: Economic happiness (y_1) Vs. 18 Causal Variables

It has been found that cropping intensity (x_9),stress (x_{18}),size of land holdings (x_3) calorie intake of family(x_{13}),education (x_2) ,media gadget cost(x_{15}), communication status (x_{11}), average family education (x_8) have been retained at the last step and has substantially contributed to the consequent variable economic happiness (y_1). These 8 variables together have contributed 79.60% of variance ie. (= 97.95 .The net contribution made by the 8 dominant causal variables stands at 97.95 percent. So the 8 variables have got formidable causal coefficient.

Stepwise Regression Analysis: Productivity happiness (y_2) Vs. 18 Causal Variables

It has been found that cropping intensity (x_9)stress (x_{18})calorie intake of family(x_{13})size of land holdings (x_3)monthly cost of smart phone(x_{17})media gadgets cost(x_{15})education (x_2) communication status (x_{11})average family education (x_8)have been retained at the last step and has substantially contributed to the consequent variable productivity happiness (y_2). These 9 variables together have contributed 73.40% of variance ie. (= 97.34 The net contribution made by the 9 dominant causal variables stands at 97.34 percent. So the 9 variables have got formidable causal coefficient.

Stepwise Regression Analysis: Family happiness (y_3) Vs. 18 Causal Variables

It has been found that stress (x_{18}) cropping intensity (x_9) food intake of family(x_{12}) size of land holdings (x_3) education (x_2) monthly cost of smartphone (x_{17}) communication status (x_{11})media gadgets cost(x_{15})have been retained at the last step and has substantially contributed to the consequent variable family happiness (y_3). these 8 variables together have contributed 77.30% of variance i.e. (= 96.98 The net contribution made by the 8 dominant causal variables stands at 96.98 percent. So the 8 variables have got formidable causal coefficient.

Stepwise Regression Analysis: neighbours happiness (y_4) Vs. 18 Causal Variables

It has been found that stress (x_{18})calorie intake of family(x_{13})size of land holdings (x_3)monthly cost of smart phone(x_{17})health status (x_{14})education (x_2)communication status (x_{11})have been retained at the last step and has substantially contributed to the consequent variable neighbours happiness (y_4). These 7 variables together have contributed 62.90% of variance i.e. (= 94.30 The net contribution made by the 7 dominant causal variables stands at 94.30 percent. So the 7 variables have got formidable causal coefficient.

Stepwise Regression Analysis: Mental health happiness (y_5) Vs. 18 Causal Variables

It has been found that media gadgets cost(x15)education ($_{x2}$)male female ratio (x7)have been retained at the last step and has substantially contributed to the consequent variable mental health happiness (y_5). These 3 variables together have contributed 62.90% of variance ie. (= 71.42 The net contribution made by the 3 dominant causal variables stands at 71.42 percent. So the 3 variables have got formidable causal coefficient.

Stepwise Regression Analysis: Weather happiness (y_6) Vs. 18 Causal Variables

It has been found that Food intake of family(x12)monthly cost of smart phone(x17)stress (x18)communication status (x11)have been retained at the last step and has substantially contributed to the consequent variable weather happiness (y_6). These 4 variables together have contributed 62.90% of variance i.e. (= 90.61 The net contribution made by the 4 dominant causal variables stands at 90.61 percent. So the 4 variables have got formidable causal coefficient.

Stepwise Regression Analysis: Ecosystem happiness (y_7) Vs. 18 Causal Variables

It has been found that cropping intensity (x_9) monthly cost of smart phone(x_{17}) media gadgets cost(x_{15})health status (x_{14}) have been retained at the last step and has substantially contributed to the consequent variable ecosystem happiness (y_7). These 4 variables together have contributed 62.90% of variance i.e. (= 82.60 The net contribution made by the 4 dominant causal variables stands at 82.60 percent. So the 4 variables have got formidable causal coefficient.

Path Analysis: Decomposition of Total Effect into Direct, Indirect and Residual Effect: Economic happiness (y_1) vs. 18 exogenous variables

It has been found that highest direct effect has been exerted by causal variable cropping intensity and the highest indirect effect has been exerted by Size of land holdings (x_3). And interestingly it has routed the highest indirect effect of 11 other causal variables to ultimately characterized economic happiness.

Path Analysis: Decomposition of Total Effect into Direct, Indirect and Residual Effect: productivity happiness (y_2) vs. 18 exogenous variables

It has been found that highest direct effect has been exerted by causal variable cropping intensity (x_9)and the highest indirect effect has been exerted by Size of land holdings (x_3). And interestingly it has routed the highest indirect effect of 10 other causal variables to ultimately characterized productivity happiness (y_2).

Path Analysis: Decomposition of Total Effect into Direct, Indirect and Residual Effect: Family happiness (y_3) vs. 18 exogenous variables

It has been found that highest direct effect has been exerted by causal variable stress (x_{18}) and the highest indirect effect has been exerted by cost of smart phone (x_{17}). And interestingly it has routed the highest indirect effect of 8 other causal variables to ultimately characterized family happiness (y_3).

Path Analysis: Decomposition of Total Effect into Direct, Indirect and Residual Effect: Neighbours happiness (y_4) vs. 18 exogenous variables

It has been found that highest direct effect has been exerted by causal variable stress (x_{18}) and the highest indirect effect has been exerted by Family food intake (x_{12}). And interestingly it has routed the highest indirect effect of 8 other causal variables to ultimately characterized neighbours happiness (y_4).

Path Analysis: Decomposition of Total Effect into Direct, Indirect and Residual Effect: Mental health happiness (y_5) vs. 18 exogenous variables

It has been found that highest direct effect has been exerted by causal variable Cost of media gadgets (x15) and the highest indirect effect has been exerted by Energy consumption (x16). And interestingly it has routed the highest indirect effect of 5 other causal variables to ultimately characterized mental health happiness (y5).

Path Analysis: Decomposition of Total Effect into Direct, Indirect and Residual Effect: Weather happiness (y_6) vs. 18 exogenous variables

It has been found that highest direct effect has been exerted by causal variable food intake (x_{12}) and the highest indirect effect has been exerted by stress (x_{18}). And interestingly it has routed the highest indirect effect of 10 other causal variables to ultimately characterized weather happiness (y_6).

Path Analysis: Decomposition of Total Effect into Direct, Indirect and Residual Effect: Ecosystem happiness (y_7) vs. 18 exogenous variables.

It has been found that highest direct effect has been exerted by causal cropping intensity (x_9) and the highest indirect effect has been exerted by irrigated land (x_4). And interestingly it has routed the highest indirect effect of 6 other causal variables to ultimately characterized ecosystem happiness (y_7).

Factor Analysis: Strategic Conglomeration of Variables into Factor.

The strategy conglomeration of operationally similar variables, called principle components based on factor loading. Factor analysis (Principle Component Analysis) is a data reduction approach. Here the total number of interacting variable, based on factor loading and Eigen values, has been grouped into identifiable factor. Each factor is retaining the homogeneous variable, strongly bound to each other towards taking a strategic decision for effective management of resource or organization.

Factor 1: Accommodation of following variables Size of land holdings (x3) Health status (x14)Monthly Cost of smart phone(x17)Economic happiness (y_1)Productivity happiness (y_2)Family happiness (y_3)Neighbours happiness

(y_4)Ecosystem happiness (y_7) has contributed 22.17 percent of variance. This factor has been renamed as ***Happiness ecology.***

Factor 2: Accommodation of following Cropping intensity (x9)Food intake of family(x12)Cost of energy consumption (x16)Stress (x18) Weather happiness(y6)has contributed 12.287 percent of variance. This factor has been renamed as ***Social ecology with happiness.***

Factor 3: Accommodation of following Irrigated land (x4) Fragments of land (x5) Family size (x6)Average family education (x8)Media gadgets cost(x15) has contributed 9.655 percent of variance. This factor has been renamed as ***Resource networking.***

Factor 4: Accommodation of following Age (x1) Education ($_{x2}$) has contributed 7.520 percent of variance. This factor has been renamed as ***Personal profile.***

Factor 5: Accommodation of following Income per year(x10) Calorie intake of family(**x13)** has contributed 6.505 percent of variance. This factor has been renamed as ***Food economics.***

Factor 6: The variable Male female ratio (x7) is solidated so the variable remains unchanged.

Factor 7: The variable Mental health happiness (y5) is solidated so the variable remains unchanged.

Micro sociological policy is being formulated for ensuring happiness to the farmers.

The CCA has helped generated by lateral strategy implication for ushering happiness among farmers under study.

7.2 Conclusion

The farmers are thriving with their livelihood among happiness, agony, certainty, uncertainty, ecstasy, melancholy; all these are contributing to their happiness or stress. More than 3 lakhs farmers have committed suicide in India, and many more at present are reeling under stress and depression. When mind is under stress, happiness is mantled by the flakes of icy depression. It is very difficult to exert the efficacy which is equally true for an individual or more for a farmer as well.

The study evinces that happiness is a binary of mental and physical well being, the duet recital of *sarod and sitar.* A pool of respondents of 100 farmers from Haringhata block of Nadia , small and marginal's by land category, has been selected and they have undergone battery of 25 questions to reveal their mind and perception regarding happiness, stress, yearnings.

The factors like cropping intensity, food and caloric intake, education, communication have been found to build up a mind set for happiness, both explicitly and implicitly.

The CCA has helped up to execute and delineate how do the two sets of variables are interactive over a space and time with reference to the variables are interactive over space and time with reference to the variables and social ecology. It has helped immensely to go for a prudent micro sociological strategy for creating an environment and condition where in a farmers could go happier.

The factors created are ***Happiness ecology, Social ecology with happiness, Resource networking, Personal profile, and Food economics*** could be extreme importance when happiness being urged upon terrains of resources, constraints, opportunities, limitations.

Farmers are providing food for every alternate hour to make us happy. They are working in torrential rain and under burning sun to save the society and the nation. If they themselves are unhappy, how could we expect the nation will go happy, else every bit of smiles hovering around our face will look like a glory of deception or something fake to hush up something wrong.

7.3 Recommendations

- The mental health of farmers and their families are to be taken care of not by chance but by a policy.
- Community mobilization approach can go a long way in reducing stress and accelerating happiness.
- The brunt of uncertainty from vagaries of weather, uncertainty of market, fragile income and cryptic livelihood should be removed with a zero tolerance.
- The factors already identified can be recommended as the techno-social input for dealing with the happiness issue.
- Indexing farmer's happiness is an essential ingredient for effective monitoring and evaluation and tracking of farmer's status of mental and physical wellbeing.
- Farmers are need to be categorized not only based on size of holdings or income, before that and beyond that threshold be a farmers category based on happiness status.

8

Future Scope

Future scope of the Research

The present study has been conducted by driving the insights deeper into the nature, intensity and character of the problems creating barrier against the farmers happiness and there mental health. The present study thus leaves behind the following domains to be researched out in future:

The study can be extrapolated to other domains of mental health and happiness:

1) Application of artificial intelligence and deep learning in the future prediction of happiness open up in vistas of research.
2) The present research can be replicated with varied geo-spatial situations viz., in different locations, settings, respondents, spatio-temporal expand and methodological innovation.
3) Policy research in the domain of happiness offers a huge scope and opportunity for the much needed extrapolation of the study.
4) Inclusion of more variables which could have been effective
5) Ecological issues and resilience should be resolved.
6) Involvement of GIS and AI in the farmers' mental health and income.
7) Social networking, business communication co-integrating communication behavior with market analysis, and data analytics in agricultural marketing and rural development studies ,can go a long way to generate alternative and scenario analysis in extension science.
8) Gender issues can be employed to a greater extent in further studies on this topic.
9) Empirical researchers are essential to estimate the value of economic and productivity losses and their toll on mental health.
10) Some new statistical tools may come up as essential and some may prove not relevant.
11) PRA materials to compensate inadequate if any in the classical statistical approaches.

Bibliography

Andrews, F. M., and Crandall, R. (1976). The validity of measures of self-reported well-being. *Soc. Indic. Res.*, **3**(1): 1-19.

Berg, A. O., Baird, M. A., Botkin, J. R., Driscoll, D. A., Fishman, P. A., Guarino, P. D., and Williams, J. K. (2009). National Institutes of Health State-of-the-Science Conference Statement: family history and improving health. *Ann.Intern. Med.*, **151**(12): 872-877.

Berry, H. L., Hogan, A., Owen, J., Rickwood, D., and Fragar, L. (2011). Climate change and farmers' mental health: risks and responses. *A.P.J.P.H*, **23**(2_suppl): 119S-132S.

Bhattacharyya, S., Burman, R. R., and Paul, S. (2019). The concept of measuring happiness and how India can go the Nordic Way. *Curr. Sci.*, **116**(1): 26-28.

Bradburn, N. M. (1969). The structure of psychological well-being.

Bren d'Amour, C., Reitsma, F., Baiocchi, G., Barthel, S., Güneralp, B., Erb, K. H., and Seto, K. C. (2017). Future urban land expansion and implications for global croplands. *P.N.A.S*, **114**(34): 8939-8944.

Chadchan, J., and Shankar, R. (2012). An analysis of urban growth trends in the post-economic reforms period in India. *Int.J. Sustain Built Environ.*, **1**(1): 36-49.

Chakraborty, A., and Acharya, S. K. (2018). Hunger poverty and voice: The impact and interrelation. In *Conference Proceedings*.

Cloutier, S., Jambeck, J., and Scott, N. (2014). The Sustainable Neighborhoods for Happiness Index (SNHI): A metric for assessing a community's sustainability and potential influence on happiness. *Ecol. Indic.*, **40**: 147-152.

Costa, P. T., McCrae, R. R., and Holland, J. L. (1984). Personality and vocational interests in an adult sample. *J. Appl. Psychol.*, **69**(3): 390.

Cox, D., Frere, M., West, S., and Wiseman, J. (2010). Developing and using local community wellbeing indicators: Learning from the experience of Community Indicators Victoria. *Aust. J. Soc. Issues.*, **45**(1): 71-88.

Dedehouanou, S. F. A., Swinnen, J., and Maertens, M. (2013). Does Contracting Make Farmers Happy? Evidence from Senegal. *Rev.Income Wealth*, **59**: S138–S160.

Diener, E. (1994). Assessing subjective well-being: Progress and opportunities. *Soc. Indic. Res.*, **31**(2): 103-157.

Diener, E., Sapyta, J. J., and Suh, E. (1998). Subjective well-being is essential to well-being. *Psychol. Inq.*, **9**(1): 33-37.

Frey, B. S., and Stutzer, A. (2002). What can economists learn from happiness research?. *J. Econ. Lit.*, **40**(2), 402-435.

Güneralp, B., Seto, K. C., and Ramachandran, M. (2013). Evidence of urban land teleconnections and impacts on hinterlands. *Curr. Opin. Env. Sust.*, **5**(5): 445-451.

Gurin, G., Veroff, J., and Feld, S. (1960). Americans view their mental health: A nationwide interview survey.

Hansra, B. S., Burman, R. R., and Roy, P. (2018). Climate change: scenario analysis and consequence in Indian agriculture. *J. Community Mobilization Sustain Dev.*, **13**(3): 613-617.

Headey, B., and Wearing, A. (1989). Personality, life events, and subjective well-being: Toward a dynamic equilibrium model. *J. Pers. Soc. Psychol*, **57**(4): 731.

Huang, J., Wu, S., and Deng, S. (2016). Relative income, relative assets, and happiness in urban China. *Soc. Indic. Res.*, **126**(3): 971-985.

Isen, A. M. (2000). Some perspectives on positive affect and self-regulation. *Psychol. Inq.*, **11**(3), 184-187.

Joseph, S., Linley, P. A., Harwood, J., Lewis, C. A., and McCollam, P. (2004). Rapid assessment of well□being: The short depression□happiness scale (SDHS). Psychology and psychotherapy: Theory, research and practice, **77**(4): 463-478.

Juster, F. T., and Stafford, F. P. (1985). Time, goods, and well-being.

Kaczmarek, L. D. (2017). Eudaimonic motivation. *Encyclopedia of personality and individual differences. New York, NJ: Springer.*

Kumar, A., Singh, R. K. P., Singh, K. M., Kumar, U., and Kumar, P. (2020). Incidence of Poverty and Livelihood Status of Farmers in Bihar: Village Level Evidences. *J. Community Mobilization Sustain Dev.,* **15**(2), 365-371.

Kumar, P., Kumar, P., and Garg, R. K. (2021). A study on farmers' satisfaction and happiness after the land sale for urban expansion in India. *Land Use Policy*, **109**, 105603.

Kurukulasuriya, P., and Rosenthal, S. (2003). Climate change in agriculture-impacts and adaptations. *World Bank Climate Change Series.*

Lee, W. S., and Zhao, Z. (2017). Height, weight and well-being for rural, urban and migrant workers in China. *Soc. Indic. Res.*, **132**(1), 117-136.

Leyden, K. M., Goldberg, A., and Michelbach, P. (2011). Understanding the pursuit of happiness in ten major cities. *Urban. Aff. Rev.*, **47**(6), 861-888.

Liebman, B. H., and Reynolds, K. M. (2006). The returns from rent□seeking: campaign contributions, firm subsidies and the Byrd Amendment. *C.J.E /Revue canadienne d'économique*, **39**(4), 1345-1369.

Liu, L., Wang, T., Xie, L., and Zhan, X. (2020). Influencing factors analysis on land-lost farmers' happiness based on the rough DEMATEL method. Discrete. Dyn. Nat Soc, 2020.

Modelmog, I. (1998). 'Nature'as a Promise of Happiness: Farmers' Wives in the Area of Ammerland, Germany. *Sociologia ruralis*, **38**(1): 109-122.

Mohan, M., and Mishra, S. (2021). India's response to the covid-19 pandemic: a frontal assault on the "historically dispossessed". *Int. J. Health Serv.*, **51**(1): 107-114.

Nettle, D. (2005). *Happiness: The science behind your smile*. OUP Oxford.

Ng, Y. K. (2015). Happiness, life satisfaction, or subjective well-being? A measurement and moral philosophical perspective. *Nanyang Technological University.*

Nguyen, D. M. (2007). The catfish antidumping in perfect and imperfect competition and the role of aquaculture in farmers' happiness. Auburn University.

Oswald, A. J., Proto, E., and Sgroi, D. (2015). Happiness and productivity. *J. Labor Econ*, **33**(4): 789-822.

Paralkar, S., Cloutier, S., Nautiyal, S., and Mitra, R. (2017). The sustainable neighborhoods for happiness (SNFH) decision tool: Assessing neighborhood level sustainability and happiness. *Ecol. Indic.*, **74**: 10-18.

Parvathamma, G. L. (2016). Farmers suicide and response of the Government in India-An Analysis. *IOSR-JEF*, **7**(3): 1-6.

Parwez, S., and Sen, V. (2016). Special economic zone, land acquisition, and impact on rural India. *Emerg. Econ. Stud*, **2**(2): 223-239.

Sadashiv, K. R. (2015). Farmers Suicide in India-Causes and Remedies: 2006-2010. *J. Econ. Sustain. Dev.*, **6**(1): 1-8.

Sareen, J., Cox, B. J., Afifi, T. O., De Graaf, R., Asmundson, G. J., Ten Have, M., and Stein, M. B. (2005). Anxiety disorders and risk for suicidal ideation and suicide attempts: a population-based longitudinal study of adults. *Arch. Gen. Psychiatry*, **62**(11): 1249-1257.

Schmitz, A., Schmitz, T. G., and Seale Jr, J. L. (2004). *The optimal processor tariff under the Byrd Amendment* (No. 1604-2016-134426).

Schwarz, N., and Strack, F. (1999). Reports of subjective well-being: Judgmental processes and their methodological implications. *Well-being: The foundations of hedonic psychology*, **7**: 61-84.

Sengupta, R. (2020). *Every day, 28 people dependent on farming die by suicide in India.*

Sharma, N., and Acharya, S. (2019). Social Entropy: Process of Technology Socialization and The Indian Agriculture.

Somerville, L. H., Kim, H., Johnstone, T., Alexander, A. L., and Whalen, P. J. (2004). Human amygdala responses during presentation of happy and neutral faces: correlations with state anxiety. *Biol. Psychiatry*, **55**(9): 897-903.

Song, Y., Zhou, A., and Zhang, M. (2020). Exploring the effect of subjective air pollution on happiness in China. *E.S.P.R*, **27**(34): 43299-43311.

Srivastava, R. (2020). Labour Migration, Vulnerability, and Development Policy: The Pandemic as Inflexion Point?. *The Indian J. Labour Econ.*, **63**(4): 859-883.

Stutzer, A., and Frey, B. S. (2008). Stress that doesn't pay: The commuting paradox. *Scand. J. Econ.*, **110**(2): 339-366.

Swami, V. (2008). Translation and validation of the Malay subjective happiness scale. *Soc. Indic. Res*, ***88***(2): 347-353.

Tang, L., Luo, X., Yu, W., and Huang, Y. (2020). The effect of political participation and village support on farmers happiness. *J.C.P.S*, **25**(4): 639-661.

Veenhoven, R. (2012). Happiness: Also known as "life satisfaction" and "subjective well-being". In *Handbook of social indicators and quality of life research* (pp. 63-77). Springer, Dordrecht.

Waage, S. A., Geiser, K., Irwin, F., Weissman, A. B., Bertolucci, M. D., Fisk, P., and McPherson, A. (2005). Fitting together the building blocks for sustainability: a revised model for integrating ecological, social, and financial factors into business decision-making. *J. Clean. Prod*, **13**(12): 1145-1163.

Walsh, L. C., Boehm, J. K., and Lyubomirsky, S. (2018). Does happiness promote career success? Revisiting the evidence. *J. Career Assess.*, **26**(2): 199-219.

Zeigler-Hill, V., and Shackelford, T. K. (Eds.) (2020). *Encyclopedia of personality and individual differences*. Cham: Springer International Publishing.

Zhang, X., Zhang, X., and Chen, X. (2017). Happiness in the air: How does a dirty sky affect mental health and subjective well-being?. *J. Environ. Econ Manag.*, **85**: 81-94.

Zhu, Z., Ma, W., Leng, C., and Nie, P. (2021). The relationship between happiness and consumption expenditure: evidence from rural China.*A.R.Q.O.L* **16**(4): 1587-1611.

Zidanšek, A. (2007). Sustainable development and happiness in nations. *Energy*, **32**(6): 891-897.